T0281531

The Elements of
Advanced Mathematics

Textbooks in Mathematics

Series editors:

Al Boggess, Kenneth H. Rosen

https://www.routledge.com/Textbooks-in-Mathematics/book-series/CANDHTEXBOOMTH

The Elements of
Advanced Mathematics
Fifth Edition

Steven G. Krantz

CRC Press
Taylor & Francis Group
Boca Raton London New York

CRC Press is an imprint of the
Taylor & Francis Group, an **informa** business
A CHAPMAN & HALL BOOK

Fifth edition published 2022
by CRC Press
6000 Broken Sound Parkway NW, Suite 300, Boca Raton, FL 33487-2742

and by CRC Press
4 Park Square, Milton Park, Abingdon, Oxon, OX14 4RN

© 2022 Steven G. Krantz

First edition published by CRC Press 1995
Second edition published by CRC Press 2002
Third edition published by CRC Press 2012
Fourth edition published by CRC Press 2017

CRC Press is an imprint of Taylor & Francis Group, LLC

ISBN: 978-1-032-10275-7 (hbk)
ISBN: 978-1-032-10279-5 (pbk)
ISBN: 978-1-003-21456-4 (ebk)

DOI: 10.1201/9781003214564

Typeset in CMR10 font
by KnowledgeWorks Global Ltd.

To the memory of R. P. Boas, 1912–1992.
For his commitment to mathematics, and for the clarity of his vision.

Contents

Preface

Overview

The character of lower division mathematics courses in universities in the United States is, and should be, different from that of upper division mathematics courses. Oversimplifying a bit, we might say that lower division courses concentrate on technique, while upper division courses treat theory.

In order to achieve any depth, an upper division mathematics course must use a precise language and methodology. The standard mathematical language includes logic, set theory, the use of functions, equivalence relations, rigorous proofs, axiomatic structures, and so forth. We frequently find ourselves, when teaching an upper division mathematics course, giving a whirlwind treatment of these basic ideas during the first week or two of the class; we also find ourselves playing catch-up during the remainder of the term.

Such a practice results in needless repetition of these common tools. It seems logical, and practical, to give the student a considered exposure to these ideas once and for all, before upper division work is commenced. That is the purpose of the present book.

Audience

Let me stress that this is not, in the strict sense, a book of logic; nor is it a book of set theory. Logicians may disapprove of my dismissal of certain subtleties. For instance, I shall neither compare the merits of various versions of set theory, nor shall I discuss attempts (such as Martin's axiom) to work around the independence of the continuum hypothesis. Rather, my purpose is to give the student, typically a first-semester or second-semester sophomore, a quick introduction to *one version* of the foundations of mathematics. In short, this is not a book for mathematicians; it is a book for students.

The student who has spent a semester studying this book should, in principle, be properly prepared for a course in real analysis or elementary Riemannian geometry or abstract algebra. Of course, there is no substitute for mathematical sophistication and hard work. This book merely provides the student with the tools of the trade.

Prerequisites for this book are minimal. Formally, the only prerequisite is the ability to read English. But, truth be told, a certain amount of exposure to mathematics and mathematical methodology is recommended. Every chapter contains a significant number of exercises. These are not merely window dressing; they are essential for mastery of the material. The student should do as many of these as possible.

When I teach this course, from this textbook of course, the students usually find it to be a life-changing experience. They feel empowered. I hope and trust that your experience will be similar.

Using This Book in Class

And now a few remarks about the layout of the book. A typical course will cover Chapters 1 through 4 rather thoroughly. These cover the basics of logic, proofs, set theory, relations, functions, and cardinality. Chapter 5 is one of the main points of the book: to construct the number systems that we use in mathematics. This material is a good venue for the student to sharpen the skills and ideas developed in earlier chapters. Some instructors may wish to skip the complex numbers, the quaternions, and the Cayley numbers, but the basic number systems should certainly be treated in detail. Chapter 6 examines the real number system more deeply than in Chapter 5. In particular, in Chapter 6 we construct and study the Cantor set.

That is a nice place to stop the course. But, if there is time, then it is a good idea to study one of the "dessert" chapters. These are Chapter 7 on number theory, Chapter 8 on cryptography, and Chapter 9 on axiomatic theories. The last chapter includes brief treatments of group theory and non-Euclidean geometry. These give the students an opportunity to see axiomatics in action.

What Is in This New Edition?

New topics include

- Further explanation of propositional logic, predicate logic, first-order logic, and related ideas. Especially in view of the prominence of theoretical computer science and its symbiotic relationship with logic, we feel that it is important to develop these ideas fully.

- Construction and discussion of the nonstandard real numbers.

- A new chapter that explores deeper properties of the real numbers. This includes topological issues and the construction of the Cantor set.

- A more exhaustive treatment of proof techniques. We have added more on induction, on counting arguments, and on enumeration and dissection. There are more geometric proofs.

- A more-thorough treatment of the Axiom of Choice and its equivalents. This includes a discussion of the Banach–Tarski paradox.

- An explicit discussion of Zorn's lemma, the Hausdorff maximality principle, and other equivalents of the Axiom of Choice.

- A treatment of partial orderings, total orderings, and well orderings. These fit rather naturally into the context of our treatment of relations.

- A fuller discussion of independence and consistency. Again, students with an interest in computer science may especially appreciate this material.

- Additional material on Russell's paradox and related ideas.

- The material on axiomatics has been moved from Chapter 5 to an Appendix. This choice was made because this content, while important, is rather recondite. The student can get a lot out of the course without studying axiomatics. But, if there is time and will, then some effort should be spent on this material.

- Additional material on group theory. Group theory is an ideal venue in which students may experience the axiomatic method for the first time, and we endeavor to make the most of it.

- A more streamlined treatment of non-Euclidean geometry. Our discussion of this topic differs from other books in the marketplace. But we feel that this material can get rather technical rather quickly, and we have endeavored to make this section of the book as slick as possible.

In this new edition, we retain the basic features of the earlier editions of the book. However, we have made strenuous efforts to make the material more accessible to neophytes. In particular, new ideas are stepladdered to give students much needed practice and experience.

A number of figures have been added. After all, mathematics is a very visual subject, and the student needs to learn to visualize the ideas.

The exercises continue to be a strong feature of the book. They are plentiful, diverse, and robust. We have marked difficult exercises with a *. There are several new exercises, and a number of the old exercises have been fleshed out to make them more meaningful to a broader cross section of students.

We have worked hard to make the page design more elegant and more appealing. A book is, after all, a tactile object that one experiences hands on.

We have beefed up the exercise sets in all the chapters. We have expanded the treatment of proofs and added some new proof techniques. Of course, errors and omissions in the existing chapters have been handled, and the text as a whole has been polished and improved.

Acknowledgments

I am grateful to Ken Rosen for contributing many insights and ideas for improving this book. As always, I thank my editor, Bob Ross, for his encouragement and support.

— Steven G. Krantz
St. Louis, Missouri

Author Bio

Steven G. Krantz is a professor of mathematics at Washington University in St. Louis. He has previously taught at UCLA, Princeton University, and Pennsylvania State University. He has written more than 130 books and more than 250 scholarly papers and is the founding editor of the *Journal of Geometric Analysis*. An AMS Fellow, Dr. Krantz has been a recipient of the Chauvenet Prize, Beckenbach Book Award, and Kemper Prize. He received a Ph.D. from Princeton University.

1

Basic Logic

1.1 Principles of Logic

Strictly speaking, our approach to logic is "intuitive" or "naïve." Whereas in ordinary conversation, these emotion-charged words may be used to downgrade the value of that which is being described, our use of these words is more technical. What is meant is that we shall prescribe in this chapter certain rules of logic that are to be followed in the rest of the book. They will be presented to you in such a way that their validity should be intuitively appealing and self-evident. We cannot *prove* these rules. The rules of logic are the point where our learning begins. A more advanced course in logic will explore other logical methods. The ones that we present here are universally accepted in mathematics and in most of science.

We shall begin with sentential logic and elementary connectives. This material is called the *propositional calculus* (to distinguish it from the predicate calculus, which will be treated later). In other words, we shall be discussing *propositions*—which are built up from atomic statements and connectives. The elementary connectives include "and," "or," "not," "if-then," and "if and only if." Each of these will have a precise meaning and will have exact relationships with the other connectives. In Section 1.8, we shall discuss the completeness of this system of elementary sentential logic, although we shall not present the *proof* of completeness (see [STO, p. 147 ff.] for a discussion of the work of Frege, Whitehead and Russell, Bernays, and Gödel in this regard).

An *atomic statement* (or *elementary statement*) is a sentence with a subject and a verb (and sometimes an object) but no connectives (and, or, not, if-then, if-and-only-if). For example,

John is good.

Mary has bread.

Ethel reads books.

are all atomic statements. We build up sentences, or propositions, from atomic statements using connectives.

After we treat the connectives, we shall consider the quantifiers "for all" and "there exists" and their relationships with the connectives from the last

DOI: 10.1201/9781003214564-1

paragraph. The quantifiers will give rise to the so-called *predicate calculus*. Connectives and quantifiers will prove to be the building blocks of all future statements in this book, indeed in all of mathematics.

1.2   Truth

In everyday conversation, people sometimes argue about whether a statement is true or not. In mathematics there is nothing to argue about. In practice, a sensible statement in mathematics is either true or false, and there is no room for opinion about this attribute. How do we determine which statements are true and which are false?

The modern methodology in mathematics works as follows:

- We *define* certain terms.

- We *assume* that these terms have certain properties or truth attributes (these assumptions are called axioms).

- We specify certain rules of logic.

Any statement that can be derived from the axioms, using the rules of logic, is understood to be true (we call such a derivation a *proof*). It is not necessarily the case that every true statement can be derived in this fashion. However, in practice this is our method for verifying that a statement is true. See Section 1.8 for a more detailed discussion of truth versus provability.

On the other hand, a statement is false if it is inconsistent with the axioms and the rules of logic. That is to say, a statement is false if the assumption that it is true leads to a contradiction. Alternatively, a statement P is false if the negation of P can be established or proved. While it is possible for a statement to be false without our being able to derive a contradiction in this fashion, in practice we establish falsity by the method of contradiction or by giving a counterexample (which is another aspect of the method of contradiction). Again, see Section 1.8 for more on falsity versus inconsistency.

The point of view being described here is special to mathematics. While it is indeed true that mathematics is used to model the world around us—in physics, engineering, and in other sciences—the subject of mathematics itself is a man-made system. Its internal coherence is guaranteed by the axiomatic method that we have just described.

It is reasonable to ask whether mathematical truth is a construct of the human mind or an immutable part of nature. For instance, is the assertion that "the area of a circle is π times the radius squared" actually a fact of nature just like Newton's inverse square law of gravitation? Our point of view is that mathematical truth is relative. The formula for the area of a circle is a logical consequence of the axioms of mathematics, nothing more. The fact that the formula seems to describe what is going on in nature is convenient and is part of what makes mathematics useful. But that aspect is something over which we as mathematicians have no control. Our concern is with the internal coherence of our logical system.

It can be asserted that a proof (a concept to be discussed and developed later in the book) is a psychological device for convincing the reader that an assertion is true. However, our view in this book is more rigid: a proof of an assertion is a sequence of applications of the rules of logic to derive the assertion from the axioms. There is no room for opinion here. The axioms are plain. The rules are rigid. A proof is like a sequence of moves in a game of chess. If the rules are followed, then the proof is correct; otherwise not.

1.3 "And" and "Or"

Let A and B be atomic statements such as "Chelsea is smart" or "The earth is flat." The statement

$$\text{``}A \quad \text{and} \quad B\text{''}$$

means that both A is true and B is true. For instance,

$$\text{Arvid is old and Arvid is fat.}$$

means both that Arvid is old *and* Arvid is fat. If we meet Arvid and he turns out to be young and fat, then the statement is false. If he is old and thin, then the statement is false. Finally, if Arvid is *both* young and thin, then the statement is false. The statement is *true* precisely when both properties—oldness and fatness—hold. We may summarize these assertions with a *truth table*. We let

$$A = \text{ Arvid is old.}$$

and

$$B = \text{ Arvid is fat.}$$

The expression

$$A \wedge B$$

will denote the phrase "A and B." We call this statement the *conjunction* of A and B. The letters "T" and "F" denote "True" and "False," respectively. Then we have

A	B	$A \wedge B$
T	T	T
T	F	F
F	T	F
F	F	F

Notice that we have listed all possible truth values of A and B and the corresponding values of the *conjunction* $A \wedge B$.

In a restaurant, the menu often contains phrases such as

soup or salad

This means that we may select soup *or* salad, but we may not select both. This use of "or" is called the *exclusive* "or"; it is not the meaning of "or" that we use in mathematics and logic. In mathematics we instead say that "A or B" is true provided that A is true or B is true or *both* are true. This is the *inclusive* "or." If we let $A \vee B$ denote "A or B," then the truth table is

A	B	$A \vee B$
T	T	T
T	F	T
F	T	T
F	F	F

We call the statement $A \vee B$ the *disjunction* of A and B.

We see from the truth table that the only way that "A or B" can be false is if *both* A is false and B is false. For instance, the statement

Hillary is beautiful or Hillary is poor.

means that Hillary is either beautiful or poor or both. In particular, she will not be both ugly and rich. Another way of saying this is that if she is ugly she will compensate by being poor; if she is rich she will compensate by being beautiful. *But she could be both beautiful and poor.*

EXAMPLE 1.3.1 The statement

$$x > 2 \quad \text{and} \quad x < 5$$

is true for the number $x = 3$ because this value of x is both greater than 2 *and* less than 5. It is false for $x = 6$ because this x value is greater than 2 but not less than 5. It is false for $x = 1$ because this x is less than 5 but not greater than 2.   ◇

EXAMPLE 1.3.2 The statement

x is odd and   x is a perfect cube.

is true for $x = 27$ because both assertions hold. It is false for $x = 7$ because this x, while odd, is not a cube. It is false for $x = 8$ because this x, while a cube, is not odd. It is false for $x = 10$ because this x is neither odd nor is it a cube.   ◇

EXAMPLE 1.3.3 The statement

$$x < 3 \text{ or } x > 6$$

is true for $x = 2$ since this x is < 3 (even though it is not > 6). It holds (that is, it is true) for $x = 9$ because this x is > 6 (even though it is not < 3). The statement fails (that is, it is false) for $x = 4$ since this x is neither < 3 nor > 6.   ◇

EXAMPLE 1.3.4 The statement

$$x > 1 \text{ or } x < 4$$

is true for every real x.   ◇

EXAMPLE 1.3.5 The statement $(A \vee B) \wedge B$ has the following truth table:

A	B	$A \vee B$	$(A \vee B) \wedge B$
T	T	T	T
T	F	T	F
F	T	T	T
F	F	F	F

◇

Notice in Example 1.3.5 that the statement $(A \vee B) \wedge B$ has the same truth values as the simpler statement B. In what follows, we shall call such pairs of statements (having the same truth values) *logically equivalent*.

The words "and" and "or" are called *connectives*: their role in sentential logic is to enable us to build up (or to connect together) pairs of atomic statements. The idea is to use very simple statements, like "Jennifer is swift" as building blocks; then we compose more complex statements from these building blocks by using connectives.

You will notice that we *always* lay out the truth values for A and B in a truth table in the same way. This makes it easier for us to compare the truth tables of different statements.

In the next two sections, we will become acquainted with the other two basic connectives "not" and "if-then."

1.4   "Not"

The statement "not A," written $\sim A$, is true whenever A is false. For example, the statement

Charles is not happily married.

is true provided the statement "Charles is happily married" is false. The truth table for $\sim A$ is as follows:

A	$\sim A$
T	F
F	T

Greater understanding is obtained by combining connectives:

EXAMPLE 1.4.1 Here is the truth table for $\sim (A \wedge B)$:

A	B	$A \wedge B$	$\sim (A \wedge B)$
T	T	T	F
T	F	F	T
F	T	F	T
F	F	F	T

◇

EXAMPLE 1.4.2 Now we look at the truth table for $(\sim A) \vee (\sim B)$:

A	B	$\sim A$	$\sim B$	$(\sim A) \vee (\sim B)$
T	T	F	F	F
T	F	F	T	T
F	T	T	F	T
F	F	T	T	T

◇

Notice that the statements $\sim (A \wedge B)$ and $(\sim A) \vee (\sim B)$ have the *same truth table*. As previously noted, such pairs of statements are called *logically equivalent*.

The logical equivalence of $\sim (A \wedge B)$ with $(\sim A) \vee (\sim B)$ makes good intuitive sense: the statement $A \wedge B$ fails precisely when either A is false *or* B is false. Since in mathematics we cannot rely on our intuition to establish facts, it is important to have the truth table technique for establishing logical equivalence. The exercise set will give you further practice with this notion.

One of the main reasons that we use the *inclusive* definition of "or" rather than the exclusive one is so that the connectives "and" and "or" have the nice relationship just discussed. It is also the case that $\sim (A \vee B)$ and $(\sim A) \wedge (\sim B)$ are logically equivalent. These logical equivalences are sometimes referred to as *de Morgan's Laws*.

1.5 "if-then"

A statement of the form "If A then B" asserts that, whenever A is true, then B is also true. This assertion (or "promise") is tested when A is true because it is then claimed that something else (namely B) is true as well. *However*, when A is false, then the statement "If A then B" *claims nothing*. Using the symbols $A \Rightarrow B$ to denote "If A then B," we obtain the following truth table:

A	B	$A \Rightarrow B$
T	T	T
T	F	F
F	T	T
F	F	T

Notice that we use here an important principle of Aristotelian logic: every sensible statement is either true or false. There is no "in between" status. When A is false, we can hardly assert that

$$A \Rightarrow B$$

is false. For $A \Rightarrow B$ asserts that "whenever A is true then B is true," and A is not true!

Put in other words, when A is false, then the statement $A \Rightarrow B$ is not tested. It therefore cannot be false. So it must be true.

EXAMPLE 1.5.1 The statement "If $2 = 4$, then Calvin Coolidge was our greatest president" is true (the antecedent is false and the conclusion may be true or false). This is the case no matter what you think of Calvin Coolidge.

The statement "If fish have hair, then chickens have lips" is true (the antecedent is false, and the conclusion is false).

The statement "If $9 > 5$, then dogs don't fly" is true (the antecedent is true, and the conclusion is true).

[Notice that the "if" part of the sentence and the "then" part of the sentence need not be related in any intuitive sense. The truth or falsity of an "if-then" statement is simply a fact about the logical values of its hypothesis and of its conclusion.]   ◇

EXAMPLE 1.5.2 The statement $A \Rightarrow B$ is logically equivalent to $(\sim A) \vee B$. The truth table for the latter is

A	B	$\sim A$	$(\sim A) \vee B$
T	T	F	T
T	F	F	F
F	T	T	T
F	F	T	T

which is the same as the truth table for $A \Rightarrow B$.   ◇

You should think for a bit to see that $(\sim A) \vee B$ *says the same thing* as $A \Rightarrow B$. To wit, assume that the statement $(\sim A) \vee B$ is true. Now suppose that A is true. Then the first half of the disjunction is false; so the second half must be true. In other words, B must be true. But that says that $A \Rightarrow B$. For the converse, assume that $A \Rightarrow B$ is true. This means that if A holds, then B must follow. But this may be rephrased as saying that if the first half of the disjunction $(\sim A) \vee B$ is false, then the second half is true. That merely affirms the disjunction. So the two statements are equivalent, i.e., they say the same thing.

Once you believe that assertion, then the truth table for $(\sim A) \vee B$ gives us another way to understand the truth table for $A \Rightarrow B$.

There are in fact infinitely many pairs of logically equivalent statements. But just a few of these equivalences are really important in practice—most others are built up from these few basic ones. Some of the other basic pairs of logically equivalent statements are explored in the exercises.

EXAMPLE 1.5.3 The statement

$$\text{If } x \text{ is negative, then } -5 \cdot x \text{ is positive.}$$

is true. For if $x < 0$, then $-5 \cdot x$ is indeed > 0; if $x \geq 0$, then the statement is unchallenged.   ◇

EXAMPLE 1.5.4 The statement

$$\text{If } (x > 0 \text{ and } x^2 < 0), \text{ then } x \geq 10.$$

is true since the hypothesis "$(x > 0$ and $x^2 < 0)$" is never true.   ◇

EXAMPLE 1.5.5 The statement

$$\text{If } x > 0, \text{ then } (x^2 < 0 \text{ or } 2x < 0).$$

is false since the conclusion "$(x^2 < 0$ or $2x < 0)$" is false whenever the hypothesis $x > 0$ is true. ◇

EXAMPLE 1.5.6 Let us construct a truth table for the statement $(A \vee (\sim B)) \Rightarrow ((\sim A) \wedge B)$.

A	B	$\sim A$	$\sim B$	$(A \vee (\sim B))$	$((\sim A) \wedge B)$	$(A \vee (\sim B)) \Rightarrow ((\sim A) \wedge B)$
T	T	F	F	T	F	F
T	F	F	T	T	F	F
F	T	T	F	F	T	T
F	F	T	T	T	F	F

◇

Notice that the statement $(A \vee (\sim B)) \Rightarrow ((\sim A) \wedge B)$ has the same truth table as $\sim (B \Rightarrow A)$. Can you comment on the logical equivalence of these two statements?

Perhaps the most commonly used logical syllogism is the following. Suppose that we know the truth of A and of $A \Rightarrow B$. We wish to conclude B. Examine the truth table for $A \Rightarrow B$. The only line in which both A is true and $A \Rightarrow B$ is true is the line in which B is true. That justifies our reasoning. In logic texts, the syllogism we are discussing is known as *modus ponendo ponens*.

In fact, *modus ponendo ponens* is quite classical terminology, going back at least to the nineteenth century. It is not commonly used today.

1.6   Contrapositive, Converse, and "Iff"

The statement

$$\text{If } A \text{ then } B$$

is the same as

$$A \Rightarrow B$$

or

$$A \text{ suffices for } B$$

or as saying

$$A \text{ only if } B$$

All these forms are encountered in practice, and you should think about them long enough to realize that they all say the same thing.

On the other hand,

$$\text{If } B \text{ then } A$$

is the same as saying

$$B \Rightarrow A$$

or

$$A \text{ is necessary for } B$$

or as saying

$$A \text{ if } B$$

We call the statement $B \Rightarrow A$ the *converse* of $A \Rightarrow B$. The converse of a statement is logically distinct from that original statement; the truth or falsity of one is independent of the truth or falsity of the other. Our examples will illustrate this point.

EXAMPLE 1.6.1 The converse of the statement

$$\text{If } x \text{ is a healthy horse, then } x \text{ has four legs.}$$

is the statement

$$\text{If } x \text{ has four legs, then } x \text{ is a healthy horse.}$$

Notice that these statements have very different meanings: the first statement is true, while the second (its converse) is false. For instance, a chair has four legs, but it is not a healthy horse.   ◇

EXAMPLE 1.6.2 The converse of the statement

$$\text{If } x > 0, \text{ then } 2x > 0.$$

is the statement

$$\text{If } 2x > 0, \text{ then } x > 0.$$

Notice that both statements are true.   ◇

EXAMPLE 1.6.3 The converse of the statement

$$\text{If } x > 0, \text{ then } x^2 > 0.$$

is the statement

$$\text{If } x^2 > 0, \text{ then } x > 0.$$

Notice that the first implication is true, while the second is false.   ◇

The statement

$$A \text{ if and only if } B$$

is a brief way of saying

If *A* then *B*.   *and*   If *B* then *A*.

We abbreviate *A* if and only if *B* as *A* ⇔ *B* or as *A* iff *B*. Here is a truth table for *A* ⇔ *B*:

A	*B*	*A* ⇒ *B*	*B* ⇒ *A*	*A* ⇔ *B*
T	T	T	T	T
T	F	F	T	F
F	T	T	F	F
F	F	T	T	T

Notice that we can say that A ⇔ B is true only when both *A* ⇒ *B* and *B* ⇒ *A* are true. An examination of the truth table reveals that *A* ⇔ *B* is true precisely when *A* and *B* are either both true or both false. Thus *A* ⇔ *B* means precisely that *A* and *B* are logically equivalent. One is true when and *only when* the other is true. One is false when and *only when* the other is false.

EXAMPLE 1.6.4 The statement

$$x > 0 \Leftrightarrow 2x > 0$$

is true. For if $x > 0$, then $2x > 0$; and if $2x > 0$, then $x > 0$.   ◇

EXAMPLE 1.6.5 The statement

$$x > 0 \Leftrightarrow x^2 > 0$$

is false. For $x > 0 \Rightarrow x^2 > 0$ is certainly true, while $x^2 > 0 \Rightarrow x > 0$ is false $((-3)^2 > 0$ but $-3 \not> 0)$.   ◇

EXAMPLE 1.6.6 The statement

$$\{\sim (A \vee B)\} \Leftrightarrow \{(\sim A) \wedge (\sim B)\} \qquad (*)$$

is true because the truth table for ∼(A ∨ B) and that for (∼ A) ∧ (∼ B) are the same. Thus they are logically equivalent: one statement is true precisely when the other is. Another way to see the truth of (∗) is to examine the truth table for the full statement:

A	*B*	∼ (*A* ∨ *B*)	(∼ *A*) ∧ (∼ *B*)	{∼ (*A* ∨ *B*)} ⇔ {(∼ *A*) ∧ (∼ *B*)}
T	T	F	F	T
T	F	F	F	T
F	T	F	F	T
F	F	T	T	T

◇

Given an implication

$$A \Rightarrow B,$$

the *contrapositive* statement is defined to be the implication

$$\sim B \Rightarrow \sim A.$$

The contrapositive is logically equivalent to the original implication, as we see by examining their truth tables:

A	B	$A \Rightarrow B$
T	T	T
T	F	F
F	T	T
F	F	T

and

A	B	$\sim A$	$\sim B$	$(\sim B) \Rightarrow (\sim A)$
T	T	F	F	T
T	F	F	T	F
F	T	T	F	T
F	F	T	T	T

EXAMPLE 1.6.7 The statement

> If it is raining, then it is cloudy.

has, as its contrapositive, the statement

> If there are no clouds, then it is not raining.

A moment's thought convinces us that these two statements say the same thing: if there are no clouds, then it could not be raining; for the presence of rain implies the presence of clouds.   ◇

The main point to keep in mind is that, given an implication $A \Rightarrow B$, its *converse* $B \Rightarrow A$ and its *contrapositive* $(\sim B) \Rightarrow (\sim A)$ are entirely different statements. The converse is distinct from, and *logically independent from*, the original statement. The contrapositive is distinct from, but *logically equivalent to*, the original statement.

Some classical treatments augment the concept of *modus ponendo ponens* with the idea of *modus tollendo tollens*. It is in fact logically equivalent to *modus ponendo ponens*. *Modus tollendo tollens* says

If $\sim B$ and $A \Rightarrow B$ then $\sim A$.

It is common to abbreviate *modus ponendo ponens* by *modus ponens* and *modus tollendo tollens* by *modus tollens*.

Modus tollens actualizes the fact that $(\sim B) \Rightarrow (\sim A)$ is logically equivalent to $A \Rightarrow B$. The first of these implications is of course the *contrapositive* of the second.

1.7 Quantifiers

The mathematical statements that we will encounter in practice will use the *connectives* "and," "or," "not," "if-then," and "iff." They will also use *quantifiers*. The two basic quantifiers are "for all" and "there exists."

EXAMPLE 1.7.1 Consider the statement

> All automobiles have wheels.

This statement makes an assertion about *all* automobiles. It is true because every automobile does have wheels.

Compare this statement with the next one:

> There exists a woman who is blonde.

This statement is of a different nature. It does not claim that all women have blonde hair—merely that there exists *at least one* woman who does. Since that is true, the statement is true.   ◇

EXAMPLE 1.7.2 Consider the statement

> All positive real numbers are integers.

This sentence asserts that something is true for all positive real numbers. It is indeed true for *some* positive numbers, such as 1 and 2 and 193. However, it is false for at least one positive number (such as $1/10$ or π), so the entire statement is false.

Here is a more extreme example:

> The square of any real number is positive.

This assertion is *almost* true—the only exception is the real number 0: $0^2 = 0$ is not positive. But it only takes one exception to falsify a "for all" statement. So the assertion is false.

This last example illustrates the principle that the negation of a "for all" statement is a "there exists" statement.   ◇

EXAMPLE 1.7.3 Look at the statement

> There exists a real number which is greater than 5.

In fact, there are lots of numbers that are greater than 5; some examples are $7, 42, 2\pi$, and $97/3$. Other numbers, such as 1, 2, and $\pi/6$, are not greater than 5. Since there is *at least one* number satisfying the statement, the assertion is true.   ◇

EXAMPLE 1.7.4 Consider the statement

> There is a man who is at least 10 feet tall.

This statement is false. To *verify* that it is false, we must demonstrate that *there does not exist a man who is at least 10 feet tall.* In other words, we must show that all men are shorter than 10 feet.

The negation of a "there exists" statement is a "for all" statement.

A somewhat different example is the sentence

> There exists a real number x which satisfies the equation
> $$x^3 - 2x^2 + 3x - 6 = 0.$$

There is in fact only one real number that satisfies the equation, and that is $x = 2$. Yet that information is sufficient to show that the statement true.   ◇

We often use the symbol $\forall$ to denote "for all" and the symbol $\exists$ to denote "there exists." The assertion

$$\forall x, \; x + 1 < x$$

claims that for every x, the number $x + 1$ is less than x. If we take our universe to be the standard real number system, then this statement is false. The assertion

$$\exists x, \; x^2 = x$$

claims that there is a number whose square equals itself. If we take our universe to be the real numbers, then the assertion is satisfied by $x = 0$ and by $x = 1$. Therefore, the assertion is true.

In all the examples of quantifiers that we have discussed thus far, we were careful to specify our *universe*. That is, "There is a woman such that ..." or "All positive real numbers are ..." or "All automobiles have ...". The quantified statement makes no sense unless we specify the universe of objects from which we are making our specification. In the discussion that follows, we will always interpret quantified statements in terms of a universe. Sometimes the universe will be explicitly specified, while other times it will be understood from context.

Quite often we will encounter $\forall$ and $\exists$ used together. The following examples are typical:

EXAMPLE 1.7.5 The statement

$$\forall x \, \exists y, \; y > x$$

claims that for any number x there is a number y that is greater than it. In the realm of the real numbers, this is true. In fact, $y = x + 1$ will always do the trick.

The statement

$$\exists y \, \forall x, \; y > x$$

has quite a different meaning from the first one. It claims that there is a y that exceeds *every* x. This is absurd. For instance, y does *not* exceed $x = y + 1$. ◇

EXAMPLE 1.7.6 The statement

$$\forall x \, \forall y, \; x^2 + y^2 \geq 0$$

is true in the realm of the real numbers: it claims that the sum of two squares is always greater than or equal to zero. [This statement happens to be *false* in the realm of the complex numbers. We shall learn about that number system later. When we interpret a logical statement, it will always be important to understand the context, or universe, in which we are working.]

The statement

$$\exists x \, \exists y, \; x + 2y = 7$$

is true in the realm of the real numbers: it claims that there exist x and y such that $x + 2y = 7$. Certainly the numbers $x = 3, y = 2$ will do the job (although there are many other choices that work as well). ◇

We conclude by noting that $\forall$ and $\exists$ are closely related. The statements

$$\forall x, \; A(x) \qquad \text{and} \qquad \sim \exists x, \; \sim A(x)$$

are logically equivalent. The first asserts that the statement $A(x)$ is true for all values of x. The second asserts that there exists no value of x for which $A(x)$ fails, which is the same thing.

Likewise, the statements

$$\exists x, B(x) \qquad \text{and} \qquad \sim \forall x, \; \sim B(x)$$

are logically equivalent. The first asserts that there is some x for which $B(x)$ is true. The second claims that it is not the case that $B(x)$ fails for every x, which is the same thing. The books [HALM] and [GIH] explore the algebraic structures inspired by these quantifiers.

The assertions

$$\forall x, A(x) \Leftrightarrow \sim \exists x, \sim A(x)$$

and

$$\exists x, B(x) \Leftrightarrow \sim \forall x, \sim B(x)$$

are commonly referred to as *de Morgan's Laws*. You should compare them with the de Morgan Laws that we discussed in Section 1.3.

It is worth noting explicitly that $\forall$ and $\exists$ do *not* commute. That is to say,

$$\forall x \exists y \,, F(x,y) \qquad \text{and} \qquad \exists y \forall x \,,\; F(x,y)$$

do *not* say the same thing. We invite you to provide a counterexample.

A "for all" statement is something like the conjunction of a very large number of simpler statements. For example, the statement

$$\text{For every nonzero integer } n, \; n^2 > 0 \,.$$

is actually an efficient way of saying that $1^2 > 0$ and $(-1)^2 > 0$ and $2^2 > 0$, etc. It is not feasible to apply truth tables to "for all" statements, and we usually do not do so.

A "there exists" statement is something like the disjunction of a very large number of statements (the word "disjunction" in the present context means an "or" statement). For example, the statement

$$\text{There exists an integer } n \text{ such that } P(n) = 2n^2 - 5n + 2 = 0 \,.$$

is actually an efficient way of saying that $P(1) = 0$ or $P(-1) = 0$ or $P(2) = 0$, etc. It is not feasible to apply truth tables to "there exist" statements, and we usually do not do so.

It is common to say that *first-order logic* consists of the connectives $\wedge$, $\vee$, $\sim$, $\Rightarrow$, $\Longleftrightarrow$, the equality symbol $=$, and the quantifiers $\forall$ and $\exists$, together with an infinite string of variables $x, y, z, \ldots, x', y', z', \ldots$ and, finally, parentheses $(\,,\,)$ to keep things readable (see [BAR, p. 7]). The word "first" here is used to distinguish the discussion from second-order and higher-order logics. In first-order logic, the quantifiers $\forall$ and $\exists$ always range over elements of the domain M of discourse. Second-order logic, by contrast, allows us to quantify over subsets of M and functions F mapping $M \times M$ into M. Third-order logic treats sets of function and more abstract constructs. The distinction among these different orders is often moot.

1.8   Truth and Provability

Let us look back at the ideas of this chapter and comment on the difference between truth and provability.

An elementary statement such as

$$A = \text{"George is tall."}$$

has a truth value assigned to it. It is either true or false. From the point of

view of mathematics, there is nothing to prove about this statement. Likewise
for the statement

$$B = \text{``Barbara is wise.''}$$

On the other hand, the statement

$$A \lor B.$$

is subject to mathematical analysis. Namely, it is true if at least one of A or
B is true. Otherwise it is false.

Any statement that is true regardless of the truth value of its individual
components is called a *tautology*. An example of a tautology is

$$B \Rightarrow (A \lor \sim A).$$

This statement is true all the time—regardless of the truth values of A and
B. Set up a truth table to satisfy yourself that this is the case.

Another example of a tautology is

$$(A \Rightarrow B) \Leftrightarrow (\sim A \lor B).$$

Again, you may verify that this is a tautology by setting up a truth table.

So we have two ways to think about whether a certain statement is valid
all the time: (i) to substitute in all possible truth values, and (ii) to prove the
statement from elementary first principles. We have seen two examples of (i).
Now let us think about method (ii).

In order to provide an example of a provable statement, we must isolate
in advance what are the syllogisms that we assume in advance to be true, and
what rules of logic are allowed. In a formal treatment of logic, such as [SUP]
or [STO], we would begin on page 1 of the book with these syllogisms and
rules of logic and then proceed rigidly, step by step. At each stage, we would
have to check which rule or syllogism is being applied. The present book is
not a formal treatment of logic. It is in fact a more intuitive approach. For
the remainder of the section, however, we lapse into the formal mode so that
we may learn more carefully to distinguish truth from provability.

First, which rules of logic do we allow? There is only one: *modus ponendo
ponens* is the only rule of logic (this is the rule that $A \Rightarrow B$ together with
A entails B—see Section 1.5). Now the other assumptions are these: for the
present discussion we take $\sim$ and $\lor$ as our only primitive connectives. Then

N1 $A \Rightarrow B$ is an abbreviation for $\sim A \lor B$.

N2 $A \land B$ is an abbreviation for $\sim (\sim A \lor \sim B)$.

Axiom 1 $(C \lor C) \Rightarrow C$

Axiom 2 $C \Rightarrow (C \lor B)$

Axiom 3 $(C \lor B) \Rightarrow (B \lor C)$

Axiom 4 $(B \Rightarrow A) \Rightarrow ([C \vee B] \Rightarrow [C \vee A])$

Notice that Axioms 1–4 are all "intuitively obvious." Any good axiom should have this feature, because we do not verify or prove axioms. The axioms are our starting place; nothing comes before the axioms. We just accept them. For example, let us think about Axiom 2: If we assume that C is true, then it is certainly the case that $C \vee B$ is true. In this way, we satisfy our intuition that Axiom 2 is a reasonable axiom. You may check the other axioms for yourself using similar reasoning.

In some more formal treatments, additional rules of logic are enunciated. The Axiom of Substitution (Axiom Schema of Replacement—see the Appendix) is also an important rule of logical reasoning. We shall say more about it later.

In a formal treatment of proof theory (see [BUS, p. 5 ff.]), we sometimes specify—in addition to *modus ponens*—a system of logical axioms that allow the inference of "self-evident" tautologies from no hypotheses. One such system is this (see [BUS]):

(i) $p \Rightarrow (q \Rightarrow p)$

(ii) $(p \Rightarrow q) \Rightarrow [(p \Rightarrow\, \sim q) \Rightarrow\, \sim p]$

(iii) $(p \Rightarrow q) \Rightarrow [(p \Rightarrow (q \Rightarrow r)) \Rightarrow (p \Rightarrow r)]$

(iv) $(\sim\sim p) \Rightarrow p$

(v) $p \Rightarrow (p \vee q)$

(vi) $(p \wedge q) \Rightarrow p$

(vii) $q \Rightarrow (p \vee q)$

(viii) $(p \wedge q) \Rightarrow q$

(ix) $(p \Rightarrow r) \Rightarrow [(q \Rightarrow r) \Rightarrow ((p \vee q) \Rightarrow r)]$

(x) $p \Rightarrow [q \Rightarrow (p \wedge q)]$

together with two axiom schemes for the quantifiers:

(xi) $A(t) \Rightarrow \exists x, A(x)$

(xii) $\forall x, A(x) \Rightarrow A(t)$

and two quantifier rules of inference:

(xiii) $[C \Rightarrow A(x)] \Rightarrow [C \Rightarrow \forall x, A(x)]$

(xiv) $[A(x) \Rightarrow C] \Rightarrow [\exists x, A(x) \Rightarrow C]$

We refer to axioms (i)–(x) as $\mathcal{F}$, in honor of Gottlob Frege (1848–1925). It is a remarkable fact that $\mathcal{F}$ is complete in the sense that any tautological statement of the propositional calculus can be proved using $\mathcal{F}$.

For the purposes of the present book, *modus ponendo ponens* will be the primary rule of reasoning. The reader can safely worry about no others. Any assertion that we assert to be *provable* must be derivable, using the logical rule *modus ponendo ponens*, from our notational conventions and these axioms. As an illustration, let us prove the statement $\sim (B \wedge \sim B)$. [Note that you can easily check this with a truth table; so it *is* a tautology. But now we want to *prove* it from (i) our *definitions*, (ii) our *axioms*, and (iii) our *rules of logic*.]

Now N2 above shows that the statement that we wish to prove is just $\sim B \vee B$. [We have used here the logical equivalence of $\sim\sim B$ and B. The details of this equivalence are left to you.] It is more natural to prove *this* statement since our axioms are formulated in terms of the connective $\vee$. Here is our proof:

(1)	$(B \vee B) \Rightarrow B$	by Axiom 1
(2)	$[(B \vee B) \Rightarrow B] \Rightarrow$	
	$([\sim B \vee [B \vee B]] \Rightarrow [\sim B \vee B])$	by Axiom 4
(3)	$([\sim B \vee [B \vee B]] \Rightarrow [\sim B \vee B])$	by modus ponendo ponens applied to (1), (2)
(4)	$(B \Rightarrow (B \vee B)) \Rightarrow (B \Rightarrow B)$	applying N1 to (3)
(5)	$B \Rightarrow (B \vee B)$	by Axiom 2
(6)	$B \Rightarrow B$	by modus ponendo ponens applied to (4), (5)
(7)	$\sim B \vee B$	applying N1 to (6)

That completes the proof.

Implicit in this last discussion is the question of why we can restrict attention to just the connectives $\sim$ and $\vee$. In fact, all the other connectives can be expressed in terms of just these two. As an instance, $A \wedge B$ is logically equivalent to $\sim (\sim A \vee \sim B)$. Likewise, $A \Rightarrow B$ is logically equivalent with $(\sim A) \vee B$. These statements can be checked with truth tables. It can also be shown that $\sim$ and $\wedge$ can be used to generate all the other connectives. Some combinations are not possible: $\vee$ and $\wedge$ *cannot* be used to form a statement that is equivalent with $\sim$. Again, you can use truth tables to confirm this assertion.

It is natural to ask, and we raised this question implicitly at the beginning of Section 1.1, whether every tautology is provable (that every provable statement is a tautology is an elementary corollary of our logical structure, or see [STO, p. 152]). That this is so is Frege's theorem. This statement is summarized by saying that elementary sentential logic is *complete*.

In fact Gödel (1906–1978) proved in 1930 that the so-called first-order predicate calculus is complete. The first-order predicate calculus is essentially the logic that we have described in the present chapter: it includes elementary connectives, the quantifiers $\forall$ and $\exists$, and statements P with one or more (but finitely many) variables $x_1, \ldots, x_k$. Thus, according to Gödel, any provable statement in this logic is true and, more profoundly, any true statement is provable. Gödel went on to construct a model for any consistent system of axioms. Interestingly, his proof requires the Axiom of Choice (Sections 5.1 and 5.2).

Gödel's more spectacular contribution to modern thought is that, in any logic that is complex enough to contain arithmetic, there are sensible statements that cannot be proved either true or false. More precisely, there are true statements that cannot be proved; and there are false statements that cannot be disproved. For example, Peano's arithmetic (Section 6.1) contains statements that cannot be proved either true or false. A rigorous discussion of this celebrated "incompleteness theorem" is beyond the scope of the present book. Suffice it to say that Gödel's proof consists of making an (infinite) list of all provable statements, enumerating with a system of "Gödel numbers", and then constructing a new statement that differs from each of these. Since the constructed statement could not be on the list, it also cannot be provable. For further discussion of Gödel's ideas, see [DAV], [NAN], [SMU].

Theoretical computer scientists have shown considerable interest in the incompleteness theorem. For a computer lang-
uage—even an expert system—can be thought of as a logical theory. Gödel's theorem says, in effect, that there will be statements formulable in any sufficiently complex language that cannot be established through a sequence of logical steps from first principles. For more on this matter, see [KAR], [SCH], [STO].

Kurt Gödel (1906–1978)

Kurt Gödel had quite a happy childhood. He had rheumatic fever when he was six years old, but after he recovered life went on much as before.

Gödel entered the University of Vienna in 1923 still without having made a definite decision whether he wanted to specialize in mathematics or theoretical physics.

Gödel completed his doctoral dissertation under Hahn's supervision in 1929, submitting a thesis proving the completeness of the first-order functional calculus. He became a member of the faculty of the University of Vienna in 1930.

Gödel is best known for his proof of "Gödel's Incompleteness Theorems." He proved fundamental results about axiomatic systems, showing in any axiomatic mathematical system there are propositions that cannot be proved or disproved within the axioms of the system.

Now 1933 was the year that Hitler came to power. At first this had no effect on Gödel's life in Vienna; he had little interest in politics. In 1934, Gödel gave a series of lectures at Princeton. However, Gödel suffered a nervous breakdown as he arrived back in Europe. He was treated by a psychiatrist and spent several months in a sanatorium.

Despite the health problems, Gödel's research was progressing well and he proved important results on the consistency of the Axiom of Choice with the other axioms of set theory in 1935.

He visited Göttingen in the summer of 1938, lecturing there on his set theory research. He returned to Vienna and married Adele Porkert in the autumn of 1938.

In 1940 Gödel arrived in the United States, becoming a US citizen in 1948 (in fact he believed he had found an inconsistency in the United States Constitution, but the judge had more sense than to listen during his interview!). He was on the faculty of the Institute for Advanced Study from 1940 until his death. One of Gödel's closest friends at Princeton was Einstein. They each had a high regard for the other and they spoke frequently.

He received the Einstein Award in 1951 and National Medal of Science in 1974. He was a member of the National Academy of Sciences of the United States, a fellow of the Royal Society, a member of the Institute of France, and a fellow of the Royal Academy.

Toward the end of his life Gödel became convinced that he was being poisoned and, refusing to eat to avoid being poisoned, essentially starved himself to death.

Exercises

1. Construct truth tables for each of the following sentences:

    (a) $(S \wedge T) \vee \sim (S \vee T)$

(b) $(S \lor T) \Rightarrow (S \land T)$

(c) $(\sim S \lor T) \Leftrightarrow \sim (S \land \sim T)$

(d) $S \Rightarrow (S \Rightarrow (S \Rightarrow (S \Rightarrow T)))$

(e) $S \Rightarrow (\sim S \Rightarrow (S \Rightarrow (\sim S \Rightarrow T)))$

(f) $[S \land (S \land (S \land T) \land T)] \lor T$

(g) $S \land (T \lor \sim S)$

(h) $(S \land \sim T) \Rightarrow (T \land \sim S)$

2. Let

$$
\begin{aligned}
S &= \text{All fish have eyelids.} \\
T &= \text{There is no justice in the world.} \\
U &= \text{I believe everything that I read.} \\
V &= \text{The moon's a balloon.}
\end{aligned}
$$

Express each of the following sentences using the letters S, T, U, V and the connectives $\lor, \land, \sim, \Rightarrow, \Leftrightarrow$. *Do not use quantifiers.*

(a) If fish have eyelids, then there is at least some justice in the world.

(b) If I believe everything that I read then either the moon's a balloon or at least some fish have no eyelids.

(c) If either the moon is not a balloon or if there is some justice in the world, then I doubt some of the things that I read.

(d) For fish to have eyelids it is necessary for the moon to be a balloon.

(e) If fish have eyelids then there is at least some justice in the world.

(f) For there to be any justice in the world it suffices for fish to have eyelids.

(g) It is not the case that either some fish have no eyelids or that I disbelieve some of what I read.

(h) In order for the moon to be a balloon it is necessary and sufficient for at least some fish to have no eyelids and for there to be some justice in the world.

(i) If the moon is not a balloon and if I do not believe all that I read then at least some fish do not have eyelids.

(j) If I do not believe some of what I read then some fish do not have eyelids.

(k) For me to disbelieve at least some of what I read it is sufficient for there to be at least some justice in the world.

3. Let

$$S = \text{All politicians are honest.}$$
$$T = \text{Some men are fools.}$$
$$U = \text{I don't have two brain cells to rub together.}$$
$$W = \text{The pie is in the sky.}$$

Translate each of the following into English sentences:

(a) $(S \wedge \sim T) \Rightarrow \sim U$

(b) $W \vee (T \wedge \sim U)$

(c) $W \Rightarrow (S \Rightarrow T)$

(d) $S \Rightarrow (S \vee U)$

(e) $(W \wedge U) \Leftrightarrow (W \vee U)$

(f) $[\sim (W \wedge \sim T)] \vee [\sim U \wedge S]$

(g) $W \vee (\sim S \Rightarrow T)$

(h) $S \Leftrightarrow (W \Rightarrow U)$

(i) $U \Leftrightarrow (W \Leftrightarrow U)$

(j) $U \Rightarrow (U \Rightarrow W)$

4. State the converse and the contrapositive of each of the following sentences. Be sure to label each.

(a) In order for it to rain it is necessary that there be clouds.

(b) In order for it to rain it is sufficient that there be clouds.

(c) If life is a bowl of cherries, then I am not in the pits.

(d) If I am not a fool, then mares eat oats.

(e) A sufficient condition for the liquidity of water is that the temperature exceed 32° Fahrenheit.

(f) A necessary condition for peace in the world is that all people disarm.

(g) What's good for the goose is good for the gander.

(h) If wishes were horses, then beggars would ride.

(i) If my grandmother had wheels, she'd be a garbage truck.

(j) If he won't play ball, then he's benched.

(k) If the Donald won't buy it, then it's not worth a dime.

(l) If Melania wants it, then the Donald will buy it.

5. Assume that the universe is the ordinary system $\mathbb{R}$ of real numbers. Which of the following sentences is true? Which is false? Give reasons for your answers.

(a) If π is rational, then the area of a circle is $E = mc^2$.

(b) If $2 + 2 = 4$, then $3/5$ is a rational number.

(c) If $2 + 2 = 5$, then $2 + 3 = 6$.

(d) If both $2 + 3 = 5$ and $2 \cdot 3 = 5$, then the world is flat.

(e) If it is not the case that $3^2 = 9$, then $4^2 = 16$.

(f) If it is not the case that $3^2 = 9$, then $4^2 = 17$.

(g) If it is not the case that $3^2 = 8$, then $4^2 = 16$.

(h) If it is not the case that $3^2 = 8$, then $4^2 = 17$.

(i) If both $3 \cdot 2 = 6$ and $4 + 4 = 8$, then $5 \cdot 5 = 20$.

(j) If both $3 \cdot 2 = 6$ and $4 + 4 = 7$, then $5 \cdot 5 = 20$.

6. For each of the following statements, formulate a logically equivalent one using only $S, T, \sim$, and $\vee$. (Of course you may use as many parentheses as you need.) *Use a truth table or other means to explain why the statements are logically equivalent.*

(a) $S \Rightarrow \sim T$

(b) $\sim S \wedge \sim T$

(c) $S \Leftrightarrow \sim T$

(d) $S \wedge (T \vee \sim S)$

(e) $(S \vee T) \Rightarrow (S \wedge T)$

(f) $(S \wedge T) \Rightarrow (S \vee T)$

(g) $(S \Rightarrow T) \vee (T \Rightarrow S)$

(h) $[\sim (S \vee T)] \Rightarrow S$

7. Redo Exercise 6, this time finding logically equivalent statements that use only $S, T, \sim$, and $\wedge$. *Give reasons for your answers.*

8. Is it possible to find a statement that is logically equivalent to $S \Rightarrow T$ but that uses only $\wedge$ and $\vee$ (and not $\sim$)? Why or why not?

9. Translate each of the following statements into symbols, connectives, and quantifiers. Your answers should contain no words. State carefully what each of your symbols stands for. [Note: Each statement is true, but you are not required to verify the truth of the statements.]

(a) The number 5 has a positive square root.

(b) There is a quadratic polynomial equation with real coefficients that has no real root.

(c) The sum of two perfect cubes is never itself a perfect cube.

(d) If $x \cdot y \neq 0$, then $x^2 + y^2 > 0$.

(e) Every positive real number has two distinct real fourth roots.

(f) If z and w are complex numbers, then $z \cdot w$ is also a complex number.

(g) The sum of two irrational numbers need not be irrational.

(h) The product of two irrational numbers need not be irrational.

(i) The sum of two rational numbers is always rational.

(j) The square of a rational number is always rational.

(k) The square root of a rational number need not be rational.

10. In each of the following statements, you should treat the real number system $\mathbb{R}$ as your universe. Translate each statement into an English sentence. Your answers should contain no symbols—only words. [Note: Each statement is true, but you are not required to verify the truth of the statements.]

(a) $\exists x, (x \in \mathbb{R} \wedge x > 0 \wedge \sim \exists y, y > 0 \wedge y^2 = x)$

(b) $\exists x \forall y, (y > x) \Rightarrow (y > 5)$

(c) $\exists x \in \mathbb{R} \exists y \in \mathbb{R}, x^2 + y^2 < 2xy$

(d) $\exists x, x > 0 \wedge x^3 < x^2$

(e) $\sim \exists x, (x \in \mathbb{R} \wedge x > 1 \wedge x > x^2)$

(f) $\exists x, \sim (x^2 > 0 \Rightarrow x > 0)$

(g) $\exists x, (x \in \mathbb{R} \wedge \sim \exists y, y^3 = x + 1)$

(h) $\exists y \forall x, x^2 + y^4 > 2$

11. Use our standard quantifiers $\forall$ and $\exists$ to translate the sentence "There are exactly five solutions to the equation $P(m) = 0$" into symbols. Now translate the sentence "The equation $Q(m) = 0$ is satisfied by all but four integers."

* 12. An island is populated by truth-tellers and liars. You cannot tell which is which just by looking at them. You meet an inhabitant of the island. What single question (with a yes/no answer) can you ask him/her that will enable you to ascertain whether this person is a truth-teller or a liar?

13. For each of the following statements, formulate an English sentence that is its negation:

(a) The set S contains at least two integers.

(b) Mares eat oats and does eat oats.

(c) I'm rough and I'm tough and I breathe fire.

(d) This town is not big enough for both of us.

(e) I will marry Fred and disappoint Irving.

(f) I cannot marry either Selma or Flo.

(g) I will pay my taxes and avoid going to jail.

(h) If I am a good boy, then I will do fine.

(i) I love everyone and everyone loves me.

(j) If you study hard, then you will do well in school.

(k) If you get caught, then you will go to jail.

(l) If you work hard, then you will succeed.

(m) If you make more than $100,000, then you pay no income tax.

14. Which of these pairs of statements is logically equivalent? Why?

(a) $A \vee \sim B$ $\sim A \Rightarrow B$

(b) $A \wedge \sim B$ $\sim A \Rightarrow \sim B$

(c) $A \vee (\sim A \wedge B)$ $\sim \left[\sim A \wedge (A \vee \sim B) \right]$

(d) $B \Rightarrow \sim A$ $A \Rightarrow (A \vee B)$

(e) $A \Leftrightarrow \sim B$ $A \Rightarrow (\sim B \vee \sim A)$

(f) $\sim (A \vee \sim B)$ $B \wedge \sim A$

(g) $\sim (A \Rightarrow \sim B)$ $B \Rightarrow A$

15. Formulate, as an English sentence (without symbols), the negation of each of the statements in Exercise 10.

16. Give a logical demonstration (i.e., a "proof") that any statement in sentential logic can be formulated using just $\sim$ and $\wedge$. That is, given any statement P there is a logically equivalent statement P' such that P' uses only the connectives $\sim$ and $\wedge$.

17. Give a logical demonstration (i.e., a "proof") that any statement in sentential logic can be formulated using just $\sim$ and $\vee$. That is, given any statement P there is a logically equivalent statement P' such that P' uses only the connectives $\sim$ and $\vee$.

18. Give a logical demonstration (i.e., a "proof") that any statement in sentential logic can be formulated using just $\sim$ and $\Rightarrow$. That is, given any statement P there is a logically equivalent statement P' such that P' uses only the connectives $\sim$ and $\Rightarrow$.

19. Use a truth table to show that *not every* logical statement can be formulated using $\vee$ and $\wedge$.

* 20. There is actually a single logical connective—different from the ones that we have studied in the present chapter—with the property that every statement in sentential logic can be formulated *using that one connective only*. Endeavor to define such a connective.

21. You are given atomic sentences A, B, C. Also you are given the connectives $\vee$, $\wedge$, and $\sim$ (one of each). Using as many parentheses as you like, how many different compound sentences can you form using these components?

22. Explain why $\forall$ is logically equivalent to $\sim \exists \sim$.

23. Explain why $\exists$ is logically equivalent to $\sim \forall \sim$.

24. Explain why $\forall\exists$ and $\exists\forall$ are *not* the same.

2

Methods of Proof

2.1 What Is a Proof?

When a chemist asserts that a substance that is subjected to heat will tend to expand, he or she verifies the assertion through experiment. It is a consequence of the *definition* of heat that heat will excite the atomic particles in the substance; it is plausible that this in turn will necessitate expansion of the substance. However, our knowledge of nature is not such that we may turn these theoretical ingredients into a categorical proof. Additional complications arise from the fact that the word "expand" requires detailed definition. Apply heat to water that is at temperature 40° F or above, and it expands—with enough heat it becomes a gas that surely fills more volume than the original water. But apply heat to a diamond, and there is no apparent "expansion"—at least not to the naked eye.

Mathematics is a less ambitious subject. In particular, it is closed. It does not reach outside itself for verification of its assertions. When we make an assertion in mathematics, we must verify it using the rules that we have laid down. That is, we verify it by applying our rules of logic to our axioms and our definitions; in other words, we construct a *proof*. Section 1.8 contains some discussion of proofs and the rules of logic.

In modern mathematics, we have discovered that there are perfectly sensible mathematical statements that in fact *cannot* be verified in this fashion, nor can they be proven false. This is a manifestation of Gödel's Incompleteness theorem: that any sufficiently complex logical system will contain such unverifiable, indeed untestable, statements (see Section 1.8). Fortunately, in practice, such statements are the exception rather than the rule. In this book, and in almost all of university-level mathematics, we concentrate on learning about statements whose truth or falsity *is* accessible by way of proof.

This chapter considers the notion of mathematical proof. We shall concentrate on the three principal types of proof: direct proof, proof by contradiction, and proof by mathematical induction. Some other miscellaneous methods of proof will be treated as well. This includes proof by enumeration, proof by exhaustion, proof by cases, proof by contraposition, and several others.

In practice, a mathematical proof may contain elements of several or all of these techniques. You will see all the basic elements here. You should be sure to master each of these proof techniques, both so that you can recognize

DOI: 10.1201/9781003214564-2

them in your reading and so that they become tools that you can use in your own work.

2.2   Where Do You Find a Proof?

This is the most fundamental question. You are studying mathematics, you get an idea, and you formulate a statement that you would like to prove. How do you come up with a proof?

In this chapter, we present three basic proof techniques and a handful of other techniques as well. But the fact is that there are many hundreds of proof techniques, and new ones being developed every day. In the twenty-first century, we would perhaps like to think that we could have a computer search for the right proof. But no such technique is known, nor does one seem to be on the horizon.

In fact you must rely on your own ingenuity, and your own experience, and your own good luck, to find a proof. And the one that you find at first may not be the optimal proof. Typically the "first" proof of a new theorem is quite awkward. Rather technical and difficult. And hard to understand. But, over time, better proofs are found. Ultimately, if there is sufficient interest and motivation, we find the proof that, as Paul Erdős would have put it, is the one in God's book.

At this point in time you are just learning how to create proofs. This book will not ask you to prove any really difficult theorems. But you still must figure out how to prove any given statement. You will do so by trying things. Pick up your pencil and calculate. Look at some examples. One of my teachers told me that the best way to prove a theorem is to first try to prove it false. After you have had some experience creating proofs, then you will develop a sense of what types of statements are proved in a particular way. But mathematics is a process. And it is your job to participate in that process and develop your technique.

2.3   What Is a Conjecture?

A conjecture is an educated guess at something that might be true. Well, anyone—no matter how ignorant and uneducated—can make guesses. But, in mathematics, a conjecture should only be formulated by a scholar with considerable experience and standing, who has proved some recognized results in the field.

There are many famous conjectures in mathematics. The Goldbach conjecture is that any even number greater than 2 can be written as the sum of two primes. For example, $4 = 2 + 2$, $6 = 3 + 3$, $8 = 3 + 5$, $10 = 7 + 3$, and so forth. This conjecture is nearly 270 years old. And we are nowhere close to proving it.

See https://mathworld.wolfram.com/GoldbachConjecture.html for more about the Goldbach conjecture.

The twin primes conjecture is that there are infinitely many twin primes, or pairs of prime numbers that differ by 2. For instance, $3, 5$ and $11, 13$ and $29, 31$ are twin primes. The conjecture is that there are infinitely many such twin primes. The twin prime conjecture is 175 years old. In 2013, Yitang Zhang proved that there is a positive integer N less than 70 million so that that there are infinitely many prime pairs with difference less than N. See also https://www.britannica.com/science/twin-prime-conjecture .

Fermat's Last Theorem is a conjecture that garnered a great deal of publicity when it was proved by Andrew Wiles and Richard Taylor in 1995. There is an interesting history to go with this problem. In 1637, Pierre de Fermat was reading the book *Arithmetica* by Diophantus and was led to consider this question: If n is an integer greater than 2, then do there exist positive integers a, b, c such that $a^n + b^n = c^n$? Of course the answer is well known to be "yes" when $n = 2$. There are infinitely many Pythagorean triples. But for $n > 2$ matters are less clear. Fermat wrote in the margin of his book that he had discovered a wonderful proof that there were no solutions whatsoever when $n > 2$, but the margin was too narrow to contain the proof. Thus Fermat never recorded his proof. So mathematicians wondered for 358 years whether Fermat was correct. In 1995, Andrew Wiles and Richard Taylor in fact proved Fermat's result. Their proof was quite long and complicated, and used many ideas from modern number theory that Fermat could not have known.

See also https://mathworld.wolfram.com/FermatsLastTheorem.html.

Andrew John Wiles
1953–

Andrew Wiles's father, Maurice Frank Wiles (1923–2005), was the Regius Professor of Divinity at the University of Oxford. When Andrew was born, Maurice Wiles was Chaplain at Ridley Hall, Cambridge.

Wiles's interest in Fermat's Last Theorem began at a young age. He said: ... I was a ten year old and one day I happened to be looking in my local public library and I found a book on maths and it told a bit about the history of this problem and I, a ten year old, could understand it. From that moment I tried to solve it myself, it was such a challenge, such a beautiful problem, this problem was Fermat's Last Theorem.

Wiles did not work on Fermat's Last Theorem for his doctorate. He said: ... the problem with working on Fermat is that you could spend years getting nothing so when I went to Cambridge my advisor John Coates was working on Iwasawa theory of elliptic curves and I started working with him ... From 1977 until 1980 Wiles was a Junior Research fellow at Clare College, Cambridge and also a Benjamin Peirce Assistant Professor at Harvard University. In 1980 he was awarded his doctorate, then spent a while at the Sonderforschungsbereich Theoretische Mathematik in Bonn. He returned to the United States near the end of 1981 to take up a post at the Institute for Advanced Study in Princeton. He was appointed a professor at Princeton the following year.

The following describes events which changed the direction of Andrew Wiles's research: ... about ten years ago, G. Frey suggested and K. Ribet proved (building on ideas of B. Mazur and J.-P. Serre) that Fermat's Last Theorem follows from the Shimura-Taniyama-Weil conjecture. This result set the stage for Wiles's work.

In fact, Wiles abandoned all his other research when he heard what had been proved and, for seven years, he concentrated solely on attempting to prove the Shimura-Taniyama-Weil conjecture, knowing that a proof of Fermat's Last Theorem then followed.

Using Mazur's deformation theory of Galois representations, recent results on Serre's conjecture on the modularity of Galois representations, and deep arithmetical properties of Hecke algebras, Wiles (with one key step due jointly to Wiles and R. Taylor) succeeded in proving that all semistable elliptic curves defined over the rational numbers are modular. Although less than the full Shimura-Taniyama-Weil conjecture, this result does imply that the elliptic curve given above is modular, thereby proving Fermat's Last Theorem.

Wiles said "... there's no other problem that will mean the same to me. I had this very rare privilege of being able to pursue in my adult life what had been my childhood dream. I know it's a rare privilege but I know if one can do this it's more rewarding than anything one can imagine."

In 2000, Andrew Wiles became "Sir Andrew Wiles" when he was made a Knight Commander of the Order of the British Empire by the Queen.

The most famous, and most important, conjecture in all of mathematics is the Riemann hypothesis, first formulated by Bernhard Riemann in 1859. Too difficult and technical to state here explcitly, we will just say that the conjecture concerns the random distribution of prime numbers. There are a great many papers written about the Riemann hypothesis and its consequences. It is a vital and central part of modern analytic number theory. But we are nowhere near proving the Riemann hypothesis.

See https://mathworld.wolfram.com/RiemannHypothesis.html for more about the Riemann hypothesis.

2.4 Direct Proof

In this section, we shall assume that you are familiar with the positive integers, or *natural numbers* (a detailed treatment of the natural numbers appears in Section 5.1). This number system $\{1, 2, 3, \dots\}$ is denoted by the symbol $\mathbb{N}$. For now we will take the elementary arithmetic properties of $\mathbb{N}$ for granted. We shall formulate various statements about natural numbers and prove them. Our methodology will emulate the discussions in earlier sections. We begin with a definition.

Definition 2.4.1 A natural number n is said to be *even* if, when it is divided by 2, there is no remainder.

Definition 2.4.2 A natural number n is said to be *odd* if, when it is divided by 2, the remainder is 1.

You may have never before considered, at this level of precision, what is the meaning of the terms "odd" or "even." But your intuition should confirm these definitions. A good definition should be precise, but it should also appeal to your heuristic idea about the concept that is being defined.

Notice that, according to these definitions, any natural number is either even or odd. For if n is any natural number, and if we divide it by 2, then the remainder will be either 0 or 1—there is no other possibility (according to the Division Algorithm—see [HER]). In the first instance, n is even; in the second, n is odd.

In what follows we will find it convenient to think of an even natural number as one having the form $2m$ for some natural number m. We will think of an odd natural number as one having the form $2k + 1$ (or sometimes $2k - 1$) for some natural number k. Check for yourself that, in the first instance, division by 2 will result in a quotient of m and a remainder of 0; in the second instance it will result in a quotient of k and a remainder of 1.

Now let us formulate a statement about the natural numbers and prove it. Following tradition, we refer to formal mathematical statements either as *theorems* or *propositions* or sometimes as *lemmas* or *corollaries*. A theorem is supposed to be an important statement that is the culmination of some development of significant ideas. A proposition is a statement of lesser intrinsic importance. Usually a lemma is of no intrinsic interest, but is needed as a step along the way to verifying a theorem or a proposition. Finally, a corollary is usually a direct consequence of a theorem or proposition. The corollary can often be a matter of considerable interest; but its proof should be a brief consequence of the the theorem or proposition in question.

Proposition 2.4.3 *The square of an even natural number is even.*

Proof: Let us begin by using what we learned in Chapter 1. We may re-formulate our statement as "If n is even, then $n \cdot n$ is even." This statement makes a promise. Refer to the definition of "even" to see what that promise is:

> If n can be written as twice a natural number, then
> $n \cdot n$ can be written as twice a natural number.

The hypothesis of the assertion is that $n = 2 \cdot m$ for some natural number m. But then
$$n^2 = n \cdot n = (2m) \cdot (2m) = 4m^2 = 2(2m^2).$$
Our calculation shows that n^2 is twice the natural number $2m^2$. So n^2 is also even.

We have shown that the hypothesis that n is twice a natural number en-tails the conclusion that n^2 is twice a natural number. In other words, if n is even, then n^2 is even. That is the end of our proof. $\qquad\square$

Remark 2.4.4 What is the role of truth tables at this point? Why did we not use a truth table to verify our proposition? One *could* think of the statement that we are proving as the conjunction of infinitely many specific statements about concrete instances of the variable n; and then we could verify each one of those statements. But such a procedure is inelegant and, more importantly, impractical.

For our purposes, the truth table *tells us what we must do to construct a proof*. The truth table for $A \Rightarrow B$ shows that, if A is false, then there is nothing to check; whereas, if A is true, then we must show that B is true. That is just what we did in the proof of Proposition 2.4.3.

Most of our theorems are "for all" statements or "there exists" statements. In practice, it is not usually possible to verify them directly by use of a truth table.

Proposition 2.4.5 *The square of an odd natural number is odd.*

Proof: We follow the paradigm laid down in the proof of the previous propo-sition.

Assume that n is odd. Then $n = 2m + 1$ for some natural number m. But then

$$n^2 = n \cdot n = (2m + 1) \cdot (2m + 1) = 4m^2 + 4m + 1 = 2(2m^2 + 2m) + 1.$$

We see that n^2 is $2m' + 1$, where $m' = 2m^2 + 2m$. In other words, according to our definition, n^2 is odd. $\qquad\square$

Both of the proofs that we have presented are examples of "direct proof." A direct proof proceeds according to the statement being proved; for instance, if we are proving a statement about a square, then we calculate that square. If we are proving a statement about a sum, then we calculate that sum. Here are some additional examples:

EXAMPLE 2.4.6 Prove that, if n is a positive integer, then the quantity $n^2 + 3n + 2$ is even.

Proof: Denote the quantity $n^2 + 3n + 2$ by K. Observe that

$$K = n^2 + 3n + 2 = (n+1)(n+2).$$

Thus K is the product of two successive integers: $n + 1$ and $n + 2$. One of those two integers must be even. So it is a multiple of 2. Therefore K itself is a multiple of 2. Hence K must be even. ◇

Proposition 2.4.7 *The sum of two odd natural numbers is even.*

Proof: Suppose that p and q are both odd natural numbers. According to the definition, we may write $p = 2r + 1$ and $q = 2s + 1$ for some natural numbers r and s. Then

$$p + q = (2r + 1) + (2s + 1) = 2r + 2s + 2 = 2(r + s + 1).$$

We have realized $p + q$ as twice the natural number $r + s + 1$. Therefore $p + q$ is even. □

Remark 2.4.8 In some subjects, such as literary criticism or philosophy, it is common to reason by analogy, or to present ideas so that they sound good. If we did mathematics solely according to what sounds good, or what appeals intuitively, then we might reason as follows: "If the sum of two odd natural numbers is even then it must be that the sum of two even natural numbers is odd." This is incorrect. For instance, 4 and 6 are each even but their sum $4 + 6 = 10$ is *not* odd.

Intuition definitely plays an important role in the development of mathematics, but all assertions in mathematics must, in the end, be proved by rigorous methods.

EXAMPLE 2.4.9 Prove that the sum of an even integer and an odd integer is odd.

Proof: An even integer e is divisible by 2, so may be written in the form $e = 2m$, where m is an integer. An odd integer o has remainder 1 when divided by 2, so may be written in the form $o = 2k + 1$, where k is an integer. The sum of these is

$$e + o = 2m + (2k + 1) = 2(m + k) + 1.$$

Thus we see that the sum of an even and an odd integer will have remainder 1 when it is divided by 2. As a result, the sum is odd. ◇

Proposition 2.4.10 *The sum of two even natural numbers is even.*

Proof: Let $p = 2r$ and $q = 2s$ both be even natural numbers. Then

$$p + q = 2r + 2s = 2(r + s).$$

We have realized $p+q$ as twice a natural number. Therefore we conclude that $p + q$ is even.   □

Proposition 2.4.11 *Let n be a natural number. Then either $n > 6$ or $n < 9$.*

Proof: If you draw a picture of a number line then you will have no trouble convincing yourself of the truth of the assertion. What we want to learn here is to organize our thoughts so that we may write down a rigorous proof.

Our discussion of the connective "or" in Section 1.3 will now come to our aid. Fix a natural number n. If $n > 6$ then the 'or' statement is true and there is nothing to prove. If $n \not> 6$, then the truth table teaches us that we must check that $n < 9$. But the statement $n \not> 6$ means that $n \leq 6$ so we have

$$n \leq 6 < 9.$$

That is what we wished to prove.   □

EXAMPLE 2.4.12 Prove that every even integer may be written as the sum of two odd integers.

Proof: Let the even integer be $K = 2m$, for m an integer. If m is odd then we write
$$K = 2m = m + m$$
and we have written K as the sum of two odd integers. If, instead, m is even, then we write
$$K = 2m = (m - 1) + (m + 1).$$
Since m is even then both $m-1$ and $m+1$ are odd. So again we have written K as the sum of two odd integers.   ◇

EXAMPLE 2.4.13 Prove the Pythagorean theorem.

Proof: The Pythagorean theorem states that $c^2 = a^2 + b^2$, where a and b are the legs of a right triangle and c is its hypotenuse. See Figure 2.1.

Consider now the arrangement of four triangles and a square shown in Figure 2.2. Each of the four triangles is a copy of the original triangle in Figure 2.1. We see that each side of the all-encompassing square is equal to c. So the area of that square is c^2. Now each of the component triangles has base a and height b. So each such triangle has area $ab/2$. And the little square

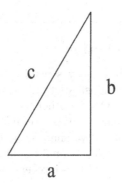

FIGURE 2.1
A right triangle.

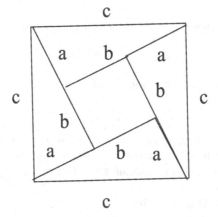

FIGURE 2.2
The Pythagorean theorem.

in the middle has side $b - a$. So it has area $(b - a)^2 = b^2 - 2ab + a^2$. We write the total area as the sum of its component areas:

$$c^2 = 4 \cdot \left[\frac{ab}{2} \right] + [b^2 - 2ab + a^2] = a^2 + b^2.$$

That is the desired equality. ◇

In this section and the next two, we are concerned with form rather than substance. We are not interested in proving anything profound, but rather in showing you what a proof looks like. Later in the book, we shall consider some deeper mathematical ideas and correspondingly more profound proofs.

2.5   Proof by Contradiction

Aristotelian logic dictates that every sensible statement has a truth value: TRUE or FALSE. If we can demonstrate that a statement A could not possibly be false, then it must be true. On the other hand, if we can demonstrate that A could not be true, then it must be false. Here is a dramatic example of this principle. In order to present it, we shall assume for the moment that you are familiar with the system $\mathbb{Q}$ of rational numbers. These are numbers that may be written as the quotient of two integers (without dividing by zero, of course).

Theorem 2.5.1 (Pythagoras) *There is no rational number x with the property that $x^2 = 2$.*

Proof: In symbols (refer to Chapter 1), our assertion may be written

$$\sim \left(\exists x, (x \in \mathbb{Q} \wedge x^2 = 2) \right).$$

Seeking a contradiction, we assume the statement to be false. Then what we are assuming is that

$$\exists x, (x \in \mathbb{Q} \wedge x^2 = 2). \tag{$*$}$$

Since x is rational, we may write $x = p/q$, where p and q are integers.

We may as well suppose that both p and q are positive and nonzero. After reducing the fraction, we may assume that it is in lowest terms—so p and q have no common factors.

Now our hypothesis asserts that

$$x^2 = 2$$

or

$$\left(\frac{p}{q} \right)^2 = 2.$$

We may write this out as

$$p^2 = 2q^2. \tag{$**$}$$

Observe that this equation asserts that p^2 is an even number. But then p must be an even number (p cannot be odd, for that would imply that p^2 is odd by Proposition 2.2.5). So $p = 2r$ for some natural number r.

Substituting this assertion into equation ($**$) now yields that

$$(2r)^2 = 2q^2.$$

Simplifying, we may rewrite our equation as

$$2r^2 = q^2.$$

This new equation asserts that q^2 is even. But then q itself must be even.

We have proven that both p and q are even. But that means that they have a common factor of 2. This contradicts our starting assumption that p and q have no common factor.

Let us pause to ascertain what we have established: the assumption that a rational square root x of 2 exists, and that it has been written in lowest terms as $x = p/q$, leads to the conclusion that p and q have a common factor and hence are *not* in lowest terms. What does this entail for our logical system?

We cannot allow a statement of the form $C = A \wedge \sim A$ (in the present context the statement A is "$x = p/q$ in lowest terms"). For such a statement C must be false.

But if x exists, then the statement C is true. No statement (such as A) can have two truth values. In other words, the statement C must be false. The only possible conclusion is that x does not exist. That is what we wished to establish. □

Remark 2.5.2 In practice, we do not include the last three paragraphs in a proof by contradiction. We provide them now because this is our first exposure to such a proof, and we want to make the reasoning absolutely clear. The point is that the assertions A and $\sim A$ cannot both be true. An assumption that leads to this eventuality cannot be valid. That is the essence of proof by contradiction.

Historically, Theorem 2.5.1 was extremely important. Prior to Pythagoras ($\sim$300 B.C.E.), the ancient Greeks (following Eudoxus) believed that all numbers (at least all numbers that arise in real life) are rational. However, by the Pythagorean theorem, the length of the diagonal of a unit square is a number whose square is 2. And our theorem asserts that such a number cannot be rational. We now know that there are many nonrational, or irrational, numbers. In fact, in Section 4.5, we shall learn that, in a certain sense to be made precise, "most" numbers are irrational.

Here is a second example of a proof by contradiction:

Theorem 2.5.3 (Dirichlet) *Suppose that $n + 1$ pieces of mail are delivered to n mailboxes. Then some mailbox contains at least two pieces of mail.*

Proof: Seeking a contradiction, we suppose that the assertion is false. Then each mailbox contains either zero or one piece of mail. But then the total amount of mail in all the mailboxes cannot exceed

$$\underbrace{1 + 1 + \cdots + 1}_{n \text{ times}}.$$

In other words, there are at most n pieces of mail. That conclusion contradicts the fact that there are $n + 1$ pieces of mail. We conclude that some mailbox contains at least two pieces of mail.   □

The last theorem, due to Gustav Lejeune Dirichlet (1805–1859), was classically known as the *Dirichletscher Schubfachschluss*. This German name translates to "Dirichlet's drawer shutting principle." Today, at least in this country, it is more commonly known as "the pigeonhole principle." Since pigeonholes are no longer a common artifact of everyday life, we have illustrated the idea using mailboxes.

EXAMPLE 2.5.4 Draw the unit interval I in the real line. Now pick 11 points at random from that interval (imagine throwing darts at the interval, or dropping ink drops on the interval). Then some pair of the points has distance not greater than 0.1 inch apart.

To see this, write

$$I = [0, 0.1] \cup [0.1, 0.2] \cup \cdots [0.8, 0.9] \cup [0.9, 1].$$

Here we have used standard interval notation. Think of each of these subintervals as a mailbox. We are delivering 11 letters (that is, the randomly selected points) to these ten mailboxes. By the pigeonhole principle, some mailbox must receive two letters.

We conclude that some subinterval of I, having length .1, contains two of the randomly selected points. Thus, their distance does not exceed 0.1 inch. ◊

EXAMPLE 2.5.5 We shall prove by contradiction that there are infinitely many prime numbers (this is an ancient result of Euclid).

Recall that a prime number is a whole number, or integer, greater than 1 which has no divisors except for 1 and itself. The first several primes are

$$2, 3, 5, 7, 11, 13, 17, 19, 23, 29, 31, \ldots .$$

A natural number which is not prime is called *composite*. A composite number will have nontrivial factors. For example, $18 = 2 \cdot 3 \cdot 3$. In particular, a composite number will always be divisible by a smaller prime.

Now, seeking a contradiction, let us suppose that there are only finitely many primes. Call them $p_1, p_2, \ldots, p_k$. Define

$$P = (p_1 \cdot p_2 \cdot p_3 \cdot \cdots \cdot p_k) + 1 .$$

What can we say about the number P?

If we divide P by p_1, then p_1 goes evenly into the product, and there is a remainder of 1. If we divide P by p_2, then p_2 goes evenly into the product, and there is a remainder of 1. And so it goes for the rest of the p_j. Now P is either prime or composite. But we just checked every known prime—p_1, $p_2, \ldots, p_k$—and verified that none of them is a divisor of P. So P cannot be composite. We conclude that P is prime.

But this is a contradiction, because P is a prime that is evidently larger than each of the p_j. We had an exhaustive list of the primes, and now we have created one more. That is a contradiction. We conclude that there are infinitely many primes. ◇

EXAMPLE 2.5.6 We shall show that there are no positive integer solutions to the equation $x^2 - y^2 = 1$. [Such an equation—a polynomial equation for which we seek integer solutions—is called a *diophantine equation*. This in honor of the ancient Greek mathematician Diophantus ($\sim$ 200 C.E.–$\sim$ 284 C.E.).]

Seeking a contradiction, we suppose that our diophantine equation *does* have integer solutions x, y. We write

$$1 = x^2 - y^2 = (x - y) \cdot (x + y).$$

Thus either both $x - y = 1$ and $x + y = 1$ or else $x - y = -1$ and $x + y = -1$. In the first case, we can add the two equations to solve them and find that $x = 1$, $y = 0$. This contradicts the assumption that both x and y are positive. In the second case, we again can add the two equations and find that $x = -1$, $y = 0$. Again, we contradict the assumption that x and y are positive.

Either case leads to a contradiction. We conclude that the diophantine equation *cannot* have a solution. ◇

EXAMPLE 2.5.7 We shall show that the sum of a rational number and an irrational number is always irrational.

Seeking a contradiction, we assume the contrary. So let q be a rational number and α an irrational number such that $q + \alpha$ is rational. So there is a rational number p with

$$q + \alpha = p.$$

But then we have

$$\alpha = p - q.$$

Surely the difference of two rational numbers is rational, so we have an irrational number equaling a rational number. That is a contradiction. ◇

2.6   Proof by Mathematical Induction

The logical validity of the method of proof by mathematical induction is intimately bound up with the construction of the natural numbers, with ordinal arithmetic, and with the so-called well-ordering principle (see Section 5.1). However, the topic fits naturally into the present chapter. So we shall present and illustrate the method, and worry about its logical foundations later on.

As with any good idea in mathematics, we shall be able to make it intuitively clear that the method is a valid and useful one. So no confusion should result.

Consider a statement $P(j)$ about the natural numbers. For example, the statement might be "The quantity $j^2 + 5j + 6$ is always even." If we wish to prove this statement, we might proceed as follows:

(1)   Prove the statement $P(1)$.

(2)   Prove that $P(j) \Rightarrow P(j+1)$ for every $j \in \{1, 2, \dots\}$.

Let us apply the syllogism *modus ponendo ponens* from the end of Section 1.5 to determine what we will have accomplished. We know $P(1)$ and, from (2) with $j = 1$, that $P(1) \Rightarrow P(2)$. We may therefore conclude $P(2)$. Now (2) with $j = 2$ says that $P(2) \Rightarrow P(3)$. We may then conclude $P(3)$. Continuing in this fashion, we may establish $P(j)$ for every natural number j.

Notice that this reasoning applies to any statement $P(j)$ for which we can establish (1) and (2) above. Thus (1) and (2) taken together constitute a method of proof. It is a method of establishing a statement $P(j)$ for every natural number j. The method is known as *proof by mathematical induction*.

It is worth enunciating the steps of the mathematical induction process in slightly different language (which should make it easier to remember):

Steps in an Inductive Proof

(a)   Enunciate the inductive statement $P(j)$. This should be a simple, declarative sentence about the positive integer j.

(b)   Verify the case $j = 1$.

(c)   Verify that the case $P(j)$ implies the case $P(j+1)$ for every $j = 1, 2, \dots$.

EXAMPLE 2.6.1 Let us use the method of mathematical induction to prove that, for every natural number j, the number $j^2 + 5j + 6$ is even.

Solution:
(a) Our statement $P(j)$ is

$$\text{The number } j^2 + 5j + 6 \text{ is even.}$$

[*Note:* Explicitly identifying $P(j)$ is more than a formality, as Exercise 2.16 shows. *Always* record carefully what $P(j)$ is before proceeding.]

We now proceed in two steps:

(b) $P(1)$ is true.   When $j = 1$ then

$$j^2 + 5j + 6 = 1^2 + 5 \cdot 1 + 6 = 12,$$

and this is certainly even. We have verified $P(1)$.

(c) $P(j) \Rightarrow P(j+1)$.   We are proving an implication in this step. We *assume* $P(j)$ and *use it* to establish $P(j + 1)$. Thus, we are assuming that

$$j^2 + 5j + 6 = 2m$$

for some natural number m. Then, to check $P(j + 1)$, we calculate

$$
\begin{aligned}
(j + 1)^2 + 5(j + 1) + 6 &= [j^2 + 2j + 1] + [5j + 5] + 6 \\
&= [j^2 + 5j + 6] + [2j + 6] \\
&= 2m + [2j + 6].
\end{aligned}
$$

Notice that, in the last step, we have *used our hypothesis* that $j^2 + 5j + 6$ is even, that is that $j^2 + 5j + 6 = 2m$. Now the last line may be rewritten as

$$2(m + j + 3).$$

Thus, we see that $(j+1)^2 + 5(j+1) + 6$ is twice the natural number $m+j+3$. In other words, $(j+1)^2 + 5(j+1) + 6$ is even. But that is the assertion $P(j+1)$.

In summary, assuming the assertion $P(j)$ we have established the assertion $P(j + 1)$. That completes Step (c) of the method of mathematical induction. We conclude that $P(j)$ is true for every j.   ◇

Here is another example to illustrate the method of mathematical induction.

Proposition 2.6.2 *If j is any natural number, then*

$$1 + 2 + \cdots + j = \frac{j(j + 1)}{2}.$$

Proof:
(a) The statement $P(j)$ is

$$1 + 2 + \cdots + j = \frac{j(j + 1)}{2}.$$

Now let us follow the method of mathematical induction closely.

(b) $P(1)$ is true. The statement $P(1)$ is

$$1 = \frac{1(1+1)}{2}.$$

This is plainly true.

(c) $P(j) \Rightarrow P(j+1)$. We are proving an implication in this step. We *assume* $P(j)$ and *use it* to establish $P(j+1)$. Thus, we are assuming that

$$1 + 2 + \cdots + j = \frac{j(j+1)}{2}. \qquad (*)$$

Let us add the quantity $(j+1)$ to both sides of $(*)$. We obtain

$$1 + 2 + \cdots + j + (j+1) = \frac{j(j+1)}{2} + (j+1).$$

The left side of this last equation is exactly the left side of $P(j+1)$ that we are trying to establish. That is the motivation for our last step.

Now the right-hand side may be rewritten as

$$\frac{j(j+1) + 2(j+1)}{2}.$$

This simplifies to

$$\frac{(j+1)(j+2)}{2}.$$

In conclusion, we have established that

$$1 + 2 + \cdots + j + (j+1) = \frac{(j+1)(j+2)}{2}.$$

This is the statement $P(j+1)$.

Assuming the validity of $P(j)$, we have established the validity of $P(j+1)$. That completes the third step of the method of mathematical induction, and establishes $P(j)$ for all j. $\qquad\qquad\square$

Some problems are formulated in such a way that it is convenient to begin the mathematical induction with some value of j other than $j = 1$. The next example illustrates this notion:

EXAMPLE 2.6.3 Let us prove that, for $j \geq 4$, we have the inequality

$$3^j > 2j^2 + 3j.$$

Solution:
(a) The statement $P(j)$ is

$$3^j > 2j^2 + 3j.$$

(b) $P(4)$ is true. Observe that the inequality is false for $n = 1, 2, 3$. However, for $n = 4$ it is certainly the case that

$$3^4 = 81 > 44 = 2 \cdot 4^2 + 3 \cdot 4.$$

(c) $P(j) \Rightarrow P(j+1)$. Now assume that $P(j)$ has been established and let us use it to prove $P(j+1)$. We are hypothesizing that

$$3^j > 2j^2 + 3j.$$

Multiplying both sides by 3 gives

$$3 \cdot 3^j > 3(2j^2 + 3j)$$

or

$$3^{j+1} > 6j^2 + 9j.$$

But now we have

$$
\begin{aligned}
3^{j+1} &> 6j^2 + 9j \\
&= 2(j^2 + 2j + j) + (4j^2 + 3j) \\
&> 2(j^2 + 2j + 1) + (3j + 3) \\
&= 2(j+1)^2 + 3(j+1).
\end{aligned}
$$

This inequality is just $P(j+1)$, as we wished to establish. That completes step three of the mathematical induction, and therefore completes the proof.

◇

EXAMPLE 2.6.4 Recall that the sequence $\{f_j\}$ of *Fibonacci numbers* is defined by

$$f_0 = 0, f_1 = 1, \ldots \quad f_j = f_{j-1} + f_{j-2} \qquad \text{for all } j \geq 2.$$

The first several Fibonacci numbers are therefore $0, 1, 1, 2, 3,$ $5, 8, 13, 21, 34, 55, 89, \ldots$. We will prove now that every third Fibonacci number is even (note that the first Fibonacci number is f_0, so we are claiming that $f_3, f_6, f_9, \ldots$ are even). We do so by mathematical induction.
(a) The statement $P(n)$ is

$$f_{3n} \text{ is even.}$$

(b) First the case $n = 1$. We may verify by inspection that $f_3 = 2$.
(c) Now suppose that the assertion has been proved for j. We will verify it for $j+1$. In fact we calculate that

$$
\begin{aligned}
f_{3(j+1)} &= f_{3j+3} \\
&= f_{3j+2} + f_{3j+1} \\
&= (f_{3j+1} + f_{3j}) + (f_{3j} + f_{3j-1}) \\
&= \left(\{f_{3j} + f_{3j-1}\} + f_{3j} \right) + (f_{3j} + f_{3j-1}) \\
&= 3f_{3j} + 2f_{3j-1}.
\end{aligned}
$$

Now we simply observe that, by the inductive hypothesis, f_{3j} is even. So the first term in the last line is even. Also, the second term has a factor of 2. So it is even. We conclude that $f_{3(j+1)}$ is even, and that completes the inductive step. ◇

EXAMPLE 2.6.5 Let us prove, by mathematical induction, that the sum of the first n positive, odd integers is equal to n^2.

We begin by writing the sum as

$$S_j = 1 + 3 + 5 + \cdots + \left[2(j-1) + 1\right].$$

Notice that when $j = 1$ this just gives 1, with $j = 2$ it gives $1 + 3$, with $j = 3$ it gives $1 + 3 + 5$, and so forth.
(a) The statement $P(j)$ is

$$S_j = j^2.$$

(b) The case $j = 1$ is the assertion $1 = 1^2$. That is certainly true.
(c) Now assume that the identity has been proved for some j. We will check it for $j + 1$. We have that

$$
\begin{aligned}
S_{j+1} &= 1 + 3 + 5 + \cdots + \left[2((j+1) - 1) + 1\right] \\
&= 1 + 3 + 5 + \cdots + \left[2j + 1\right] \\
&= \left\{ 1 + 3 + 5 + \cdots + \left[2(j-1) + 1\right] \right\} + \left[2j + 1\right] \\
&= j^2 + \left[2j + 1\right] \\
&= (j+1)^2.
\end{aligned}
$$

In the penultimate equality, we have used the inductive hypothesis. That completes the inductive step, and hence the proof. ◇

EXAMPLE 2.6.6 We present now an incorrect application of mathematical induction, just for its instructional value. In fact we will prove that

All horses are the same color.

Our statement $P(j)$ is

If S_j is a set with j horses then every
horse in S_j is of the same color.

Of course $P(1)$ is true because one horse is of the same color. Now assume that $P(j)$ is true. Set T be a set with $j + 1$ horses. Now remove one element from T. The resulting set T' has only j horses. By the inductive hypothesis, they all have the same color. Next replace that removed horse and remove a different horse. The resulting set of j horses will, by the inductive hypothesis, have the same color. We may repeat this process j times, each time removing one horse.

The conclusion is that all the horses in T have the same color. That finishes the inductive argument, and the proof is complete.

Since we all know that it is not true that all horses are the same color, there must be an error in the proof that we have just presented. What is it? The mistake is that the inductive reasoning does not work when we pass from $j = 1$ to $j = 2$. For if you have a set T with two horses, and you remove one horse, then certainly the remaining horse is of the same color. Now if you replace that horse and remove the other horse, then the remaining horse is of the same color. But so what? The two horses will, most likely, be of different colors! ◇

We conclude this section by mentioning an alternative form of the mathematical induction paradigm which is sometimes called *strong mathematical induction*.

Strong Mathematical Induction:

Let P be a function on the natural numbers. The steps are these.

(a) State $P(j)$;

(a) Prove $P(1)$;

(b) Prove that if $[P(\ell)$ is true for all $\ell \leq j]$ then $P(j+1)$ for every natural number j.

Then $P(j)$ is true for every j.

It turns out that the strong mathematical induction principle is logically equivalent to the ordinary mathematical induction principle enunciated at the outset of this section. But in some instances strong mathematical induction is the more useful tool. An alternative terminology for strong mathematical induction is "the set formulation of mathematical induction."

Strong mathematical induction is sometimes more convenient, or more natural, to use than ordinary mathematical induction; it finds particular use in abstract algebra.

EXAMPLE 2.6.7 Let us show, using strong mathematical induction, that every integer greater than 1 is either prime or the product of primes. [Here a prime number is an integer greater than 1 whose only factors are 1 and itself.]

For convenience we begin the mathematical induction process at the index 2 rather than at 1.

(a) Let $P(j)$ be the assertion "Either n is prime or n is the product of primes."

(b) Then $P(2)$ is plainly true since 2 is the first prime.

(c) Now assume that $P(\ell)$ is true for $2 \leq \ell \leq j$ and consider $P(j+1)$. If $j+1$ is prime, then we are done. If $j+1$ is not prime, then $j+1$ factors as $j+1 = k \cdot m$, where k, m are integers less than $j+1$, but at least 2. By the strong inductive hypothesis, each of k and m factors as a product of primes (or is itself a prime). Thus $j + 1$ factors as a product of primes.

The strong mathematical induction is done, and the proof is complete. $\diamond$

2.7   Other Methods of Proof

We give here a number of examples that illustrate proof techniques other than direct proof, proof by contradiction, and mathematical induction.

2.7.1   Proof by cases

One obvious but powerful method for constructing a proof is to divide the question into *cases*. We illustrate with some examples.

EXAMPLE 2.7.1 Let us show that every integer n which is a perfect cube is either (i) a multiple of 9, (ii) one less than a multiple of 9, or (iii) one more than a multiple of 9.

We consider three cases:

1.   The integer n is the cube of a multiple of 3. So $n = (3j)^3 = 27j^3 = 9 \cdot (3j^3)$. In this case, it is clear that n is a multiple of 9.

2.   The integer n is the cube of one less than a multiple of 3. Hence $n = (3j - 1)^3 = 27j^3 - 27j^2 + 9j - 1 = 9(3j^3 - 3j^2 + j) - 1$. So we see that n is one less than a multiple of 9.

3.   The integer n is the cube of one more than a multiple of 3. Therefore $n = (3j + 1)^3 = 27j^3 + 27j^2 + 9j + 1 = 9(3j^3 + 3j^2 + j) + 1$. It is clear that n is one more than a multiple of 9.

The three cases we have considered exhaust all the possibilities. So our proof is complete. $\diamond$

EXAMPLE 2.7.2 We shall show that, if n is an integer, then

$$3n^2 + n + 14$$

is even. We divide into cases.

1. If n is even, then $n = 2k$ for some integer k. Thus

$$
\begin{aligned}
3n^2 + n + 14 &= 3(2k)^2 + (2k) + 14 \\
&= 12k^2 + 2k + 14 \\
&= 2(6k^2 + k + 7).
\end{aligned}
$$

We see immediately that $3n^2 + n + 14$ is even.

2. If n is odd, then $n = 2k + 1$ for some integer k. Thus

$$
\begin{aligned}
3n^2 + n + 14 &= 3(2k+1)^2 + (2k+1) + 14 \\
&= 3(4k^2 + 4k + 1) + (2k+1) + 14 \\
&= 12k^2 + 12k + 3 + 2k + 1 + 14 \\
&= 12k^2 + 14k + 18 \\
&= 2(6k^2 + 7k + 9).
\end{aligned}
$$

We see immediately that $3n^2 + n + 14$ is even.

In both cases, which are exhaustive, our expression $3n^2 + n + 14$ is even. That completes the proof. ◇

EXAMPLE 2.7.3 We now verify the triangle inequality, which says that

$$|x + y| \leq |x| + |y|$$

for any real numbers x and y. [We note here that $|a| \leq \alpha$ if and only if $-\alpha \leq a \leq \alpha$. This fact will prove useful in the reasoning below.] To do so we divide the argument into cases:

Case 1: $x \geq 0$ and $y \geq 0$. In this case

$$|x + y| = x + y = |x| + |y|.$$

So our assertion is obvious.

Case 2: $x \geq 0$ and $y < 0$. In this case, we have

$$x + y \leq x + (-y) = |x| + |-y| = |x| + |y|. \qquad (*)$$

Furthermore,

$$-(x + y) = -x + (-y) \leq x + (-y) = |x| + |-y| = |x| + |y|.$$

Hence

$$x + y \geq -(|x| + |y|). \qquad (**)$$

Combining $(*)$ and $(**)$ now gives

$$|x + y| \leq |x| + |y|.$$

Case 3: $x < 0$ and $y \geq 0$. This case is identical to Case 2 (with the roles of x and y switched), and we omit the details.

Case 4: $x < 0$ and $y < 0$. Then

$$x + y \leq (-x) + (-y) = |x| + |y|.$$

Also

$$-(x + y) = (-x) + (-y) = |x| + |y|.$$

In conclusion,

$$|x + y| \leq |x| + |y|.$$

Combining all the cases—which certainly cover all the possibilities—we conclude that

$$|x + y| \leq |x| + |y| \qquad \text{for all } x \text{ and } y.$$

$\diamond$

EXAMPLE 2.7.4 Let us show that there exist irrational numbers a and b such that a^b is rational.

Let $\alpha = \sqrt{2}$ and $\beta = \sqrt{2}$. If α^β is rational, then we are done, using $a = \alpha$ and $b = \beta$. If α^β is irrational, then observe that

$$\left(\alpha^\beta\right)^{\sqrt{2}} = \alpha^{[\beta \cdot \sqrt{2}]} = \alpha^2 = [\sqrt{2}]^2 = 2.$$

Thus, with $a = \alpha^\beta$ and $b = \sqrt{2}$ we have found two irrational numbers a, b such that $a^b = 2$ is rational. $\diamond$

2.7.2   Proof by contraposition

Proof by contraposition is closely related to proof by contradiction. But there is a subtle difference. If we are trying to prove that $P \Rightarrow Q$ by contradiction, then we deny Q but it is unclear what contradiction we are seeking. If we instead attempt a proof by contraposition, then we seek to prove $\sim Q \Rightarrow \sim P$. We assume $\sim Q$ and our goal is to prove $\sim P$. We illustrate with some examples.

EXAMPLE 2.7.5 Let us show that, if n is an integer and n^2 is even, then n is even. We do so by contraposition. So suppose that n is odd. We shall then show that n^2 is odd.

The hypothesis then is that $n = 2k + 1$ for some integer k. Then

$$n^2 = (2k + 1)^2 = 4k^2 + 4k + 1 = 2(2k^2 + 2k) + 1.$$

Thus we see explicitly that n^2 is odd, and our result is proved. $\diamond$

EXAMPLE 2.7.6 We show that, if x, y are integers and $x + y$ is even then x and y have the same parity (i.e., either both are even or both are odd). We do so by proof by contraposition.

So suppose that x and y do *not* have the same parity. So one is even and one is odd. Say that x is even, so $x = 2j$ for some integer j, and y is odd, so $y = 2m + 1$ for some integer m. Then $x + y = (2j) + (2m + 1) = 2(j + m) + 1$. In conclusion, $x + y$ is odd. That gives the result. ⋄

EXAMPLE 2.7.7 Let us prove that, if k is an integer and $3k + 1$ is even, then k is odd. We do so by contraposition. Assume that k is even. So $k = 2m$ for some integer m. Then $3k + 1 = 3(2m) + 1 = 2(3m) + 1$, which is clearly odd. That gives the result. ⋄

2.7.3 Counting arguments

EXAMPLE 2.7.8 Show that if there are 23 people in a room, then the odds are better than even that two of them have the same birthday.

Proof: Here by "same birthday" we mean birthday on the same day of the year. For convenience we shall assume that a year has 365 days.

The best strategy is to calculate the odds that *no two* of the people have the same birthday, and then to take complements.

Let us label the people $p_1, p_2, \ldots, p_{23}$. Then, assuming that none of the p_j have the same birthday, we see that p_1 can have his/her birthday on any of the 365 days in the year, p_2 can then have his/her birthday on any of the remaining 364 days, p_3 can have his/her birthday on any of the remaining 363 days, and so forth. So the number of different ways that these 23 people can all have different birthdays is

$$365 \cdot 364 \cdot 363 \cdots 345 \cdot 344 \cdot 343.$$

On the other hand, the number of ways that birthdays could be distributed (with no restrictions) among 23 people is

$$\underbrace{365 \cdot 365 \cdot 365 \cdots 365}_{23 \text{ times}} = 365^{23}.$$

Thus the probability that these 23 people all have different birthdays is

$$p = \frac{365 \cdot 364 \cdot 363 \cdots 343}{365^{23}}.$$

A quick calculation with a pocket calculator shows that $p \sim 0.4927 < .5$. We see that the odds that 23 people will all have different birthdays is 0.4927. Thus the odds that at least two of them *will* have the same birthday is 0.5073, which is greater than one half. That is the desired result. ⋄

EXAMPLE 2.7.9 Show that if there are six people in a room, then either three of them know each other or three of them do not know each other. [Here three people know each other if each of the three pairs has met. Three people do not know each other if each of the three pairs has *not* met.]

Proof: The tedious way to do this problem is to write out all possible "acquaintance assignments" for six people. That would take a good deal of time and effort, and would be woefully inelegant.

We now describe a more efficient, and more satisfying, strategy. Call one of the people Bob. There are five others. Either Bob knows three of them, or he does not know three of them.

Say that Bob knows three of the others. If any two of those three are acquainted, then those two and Bob form a mutually acquainted threesome. If no two of those three know each other, then those three are a mutually unacquainted threesome.

Now suppose that Bob does not know three of the others. If any two of those three are unacquainted, then those two and Bob form an unacquainted threesome. If all pairs among the three are instead acquainted, then those three form a mutually acquainted threesome.

We have covered all possibilities, and in every instance come up either with a mutually acquainted threesome or a mutually unacquainted threesome. That ends the proof.   ◇

It may be worth knowing that five people is insufficient to guarantee either a mutually acquainted threesome or a mutually unacquainted threesome. We leave it to the reader to provide a suitable counterexample. It is quite difficult to determine the minimal number of people to solve the problem when "threesome" is replaced by "foursome." When "foursome" is replaced by five people, the problem is considered to be grossly intractable. This problem is a simple example from the mathematical subject known as Ramsey theory (see [GRS]).

EXAMPLE 2.7.10 Jill is dealt a poker hand of five cards from a standard deck of 52. What is the probability that she holds four of a kind?

Remark 2.7.11 In order to solve this problem we need to note the following. If you want to choose k objects from among n objects, then there are n ways to choose the first object, $(n-1)$ ways to choose the second object, $(n-2)$ ways to choose the third object, down to $(n-k+1)$ ways to choose the kth object. And these objects can be chosen in any order. There are $k! = k \cdot (k-1) \cdot (k-2) \cdots 3 \cdot 2 \cdot 1$ ways to order k objects. In conclusion, the number of ways to select k objects from among n is

$$\frac{n \cdot (n-1) \cdot (n-2) \cdots (n-k+1)}{k \cdot (k-1) \cdot (k-2) \cdots 3 \cdot 2 \cdot 1} = \frac{n!}{k!(n-k)!}.$$

This fraction occurs so frequently in mathematics that it is denoted by the special symbol

$$\binom{n}{k}.$$

Proof: If the hand holds four aces, then the fifth card is any one of the other 48 cards. If the hand holds four kings, then the fifth card is any one of the other 48 cards. And so forth. So there are a total of

$$13 \times 48 = 624$$

possible hands with four of a kind. The total number of possible five-card hands is

$$\binom{52}{5} = 2598960.$$

Here we use the standard notation $\binom{n}{k}$ to denote the number of ways to choose k objects from among n. It is known that $\binom{n}{k} = \frac{n!}{k!(n-k)!}$.

Therefore the probability of holding four of a kind is

$$p = \frac{624}{2598960} = 0.00024.$$

◇

Exercises

1. Prove that the product of two odd natural numbers must be odd.

2. Prove that if n is an even natural number and if m is *any* natural number, then $n \cdot m$ must be even.

3. Prove that the sum of the squares of the first n natural numbers is equal to
$$\frac{2n^3 + 3n^2 + n}{6}.$$

4. Prove that the sum of the first k even natural numbers is $k^2 + k$.

5. Prove—*not* by mathematical induction—that the sum of the first k odd natural numbers is k^2.

6. Prove that if n red letters and n blue letters are distributed among n mailboxes, then either some mailbox contains at least two red letters or some mailbox contains at least two blue letters or else some mailbox contains at least one red and one blue letter.

7. Prove that, if m is a power of 3 and n is a power of 3, then $m + n$ is never a power of 3.

8. What is special about the number 3 in Exercise 7? What other natural numbers can be used in its place?

9. Imitate the proof of Pythagoras's theorem to show that the number 5 does not have a rational square root.

10. Prove that if n is a natural number and if n has a rational square root, then in fact the square root of n is an integer.

11. Complete this sketch to obtain an alternative proof that the number 2 does not have a rational square root:

    (a) Take it for granted that it is known that each positive integer has one and only one factorization into prime factors (a prime number is a positive integer, greater than 1, that can be divided evenly only by 1 and itself).

    (b) Seeking a contradiction, suppose that $\alpha = p/q$ is a rational square root of 2 (we need *not* assume that the rational fraction p/q is reduced to lowest terms).

    (c) Then $2 = p^2/q^2$ or $2q^2 = p^2$.

    (d) Count the number of prime factors on either side of the last equation in part **(c)** to arrive at a contradiction.

12. Prove that if the natural number n is a perfect square, then $n + 1$ will never be a perfect square.

13. Prove that if the product of two integers is even, then one of them must be even.

14. Prove that if the product of two integers is odd, then both of them must be odd.

15. Prove that any integer can be written as the sum of at most two odd integers. Is the same true if "odd" is replaced by "even"?

* 16. A popular recreational puzzle hypothesizes that you have nine pearls that are identical in appearance. Eight of these pearls have the same weight, but the ninth is either heavier or lighter—you do not know which. You have a balance scale (see Figure 2.3), and are allowed three weighings to find the odd pearl. How do you proceed?

    Now here is a bogus proof by mathematical induction that you can solve the problem in the first paragraph in three weighings not just for nine pearls but for *any number of pearls*. For convenience let us begin the mathematical induction with the case $n = 9$ pearls. By the result of the first paragraph, we can handle that case. Now, inductively, suppose that we have an algorithm for handling j pearls. We use this hypothesis to treat $(j + 1)$ pearls. From the $(j + 1)$ pearls, remove one and put it in your pocket. There remain

FIGURE 2.3
A balance scale.

j pearls. Apply the j-pearl algorithm to these remaining pearls. If you find the odd pearl, then you are done. If you do not find the odd pearl, then it is the one in your pocket. That completes the case $(j+1)$ and the proof.

What is the flaw in this reasoning? [*Remark*: If you are fiendishly clever, then you can actually handle 12 pearls in the original problem—with just three weighings. However, this requires the consideration of 27 cases.]

In each of Exercises 17–26, either prove that the statement is true or give a counterexample. Remember that a counterexample to a "for all" statement consists of a single example; but a counterexample to a "there exists" statement consists in showing that something never occurs.

17. The sum of two perfect squares is a perfect square.

* **18.** Let n be a positive integer. In the list $n, n+1, n+2, \ldots, 2n+2$ there must be a perfect square.

19. There is a positive integer that is the sum of all its divisors that are less than itself (including the divisor 1). Such a number is called a *perfect number*.

20. The difference of two perfect squares is never a prime (refer to Exercise 11 for the definition of "prime").

21. The sum of two perfect squares is never a prime.

22. For x a positive real number we have $1 + x^2 < (1+x)^2$.

* **23.** For any positive real numbers $a_1, a_2, \ldots, a_n$ we have

$$\frac{a_1 + a_2 + \cdots + a_n}{n} \leq \left(a_1 \cdot a_2 \cdots a_n\right)^{1/n}.$$

24. Between any two distinct real numbers there is a rational number.

25. Between any two distinct rational numbers there is an irrational number.

* **26.** Let m and n be two successive perfect cubes. Then between them must lie a perfect square.

27. Prove by mathematical induction that every natural number greater than 1 has a prime factor (refer to Exercise 11 for the definition of "prime").

* **28.** Prove by mathematical induction that the sum of the angles interior to a convex polygon with k sides is $(k-2) \cdot 180°$ (begin with $k = 3$ and you may assume that the result is known for triangles).

29. Prove that if k is a natural number that is greater than 2, then $2^k > 1 + 2k$.

* **30.** Prove that the method of complete mathematical induction is logically equivalent to the method of ordinary mathematical induction.

31. Give a formal discussion of why the mathematical induction process may be begun at any natural number—not just 1.

* **32.** Fix a number $q > 0$. Use mathematical induction on the positive integer n to prove the following formula of Ramanujan:

$$1 + \frac{q}{1-q} + \frac{q^2}{(1-q)(1-q^2)} + \frac{q^3}{(1-q)(1-q^2)(1-q^3)}$$
$$+ \cdots + \frac{q^n}{(1-q)(1-q^2)\cdots(1-q^n)}$$
$$= \frac{1}{(1-q)(1-q^2)\cdots(1-q^n)}$$

* **33.** Use mathematical induction on n to prove that the formula in Exercise 32 still holds if the sequence of exponents $1, 2, 3, \ldots, n$ is replaced by the sequence of square exponents $1, 4, 9, \ldots, n^2$.

34. Prove that, if n is an integer greater than 4, then $2^n > n^2 + 1$.

35. Prove the pigeonhole principle by mathematical induction.

36. Give a direct proof of the pigeonhole principle.

37. You write 27 letters to 27 different people. Then you address the 27 envelopes. You close your eyes and stuff one letter into each envelope. What is the probability that just one letter is in the wrong envelope?

* **38.** You have a finite collection of points in the plane, not all colinear. Prove that there is a line passing through just two of these points.

* **39.** A tangled up piece of string of length 1 lies in the plane. Prove that there is a rectangle containing the string that has area not exceeding $1/4$.

40. Formulate a statement, using only the quantifier $\exists$ and *not* the quantifier $\forall$, that expresses the thought that all boys under the age of 10 practice all pieces in their piano book every day.

* **41.** Let a be a nonzero real number. Prove by mathematical induction that, for any positive integer n,

$$\begin{pmatrix} a & 2 \\ 0 & a \end{pmatrix}^n = \begin{pmatrix} a^n & 2na^{n-1} \\ 0 & a^n \end{pmatrix}$$

42. Use mathematical induction to prove that every positive integer of the form $n^3 - n$ is divisible by 6.

* **43.** Use mathematical induction to prove the identity

$$\frac{2^2}{1 \cdot 3} \cdot \frac{3^2}{2 \cdot 4} \cdots \frac{n^2}{(n-1)(n+1)} = \frac{2n}{n+1}.$$

44. Use mathematical induction to prove that

$$\frac{1}{\sqrt{1}} + \frac{1}{\sqrt{2}} + \cdots + \frac{1}{\sqrt{1998}} \geq \sqrt{1998}.$$

[**Hint:** Formulate a statement that depends on n (instead of on 1998) and prove *that* by mathematical induction on n.]

45. Imitate an example from the text to give an erroneous proof that all numbers in a set of numbers with n elements are equal.

3

Set Theory

3.1 Undefinable Terms

Even the most elementary considerations in logic may lead to conundrums. Suppose that we wish to define the notion of "line." We might say that it is the shortest path between two points. This is not completely satisfactory because we have not yet defined "path" or "point." And when we say "the shortest path," do we mean that there is just one unique shortest path? And why does it exist? Every new definition is, perforce, formulated in terms of other ideas. And those ideas in terms of other ones. Where does the regression cease?

The accepted method for dealing with this problem is to begin with certain terms (as few as possible) that are agreed to be "undefinable." These terms should be so simple that there can be little argument as to their meaning. But it is agreed in advance that these undefinable terms simply cannot be defined in terms of ideas that have been previously defined. Our undefined terms are our starting place.

In modern mathematics it is customary to use "set" and "element of" as undefinables. A *set* is declared to be a collection of objects. (Please do not ask what an "object" is or what a "collection" is; when we say that the term "set" is an undefinable, then we mean just that.) If S is a set, then we say that x is *an element of* S, and we write $x \in S$ or $S \ni x$, precisely when x is one of the objects that compose the set S. For example, we write $5 \in \mathbb{N}$ to indicate that the number 5 is an element of the set of natural numbers. We write $-7 \notin \mathbb{N}$ to specify that -7 is *not* an element of the set of natural numbers.

Definition 3.1.1 We say that two sets S and T are equal precisely when they have the same elements. We write $S = T$.

As an example of equality of sets, if $S = \{x \in \mathbb{N} : x^2 > 3\}$ and $T = \{x \in \mathbb{N} : x \geq 2\}$, then $S = T$.

Incidentally, the method of specifying a set with the notation $\{x : P(x)\}$, where P denotes a property, is the most common method in mathematics of defining a set. This is sometimes called "setbuilder notation." When we discuss the formal axiomatics of set theory in the Appendix, we will see that one of the axioms explicitly mandates that sets may be described in this way.

DOI: 10.1201/9781003214564-3

We shall endeavor, in what follows, to formulate all of our set-theoretic notions in a rigorous and logical fashion from the undefinables "set" and "element of." If at any point we arrive at an untenable position, or a logical contradiction, or a fallacy, then we know that the fault lies with either our method of reasoning or with our undefinables or with our axioms. One of the advantages of the way that we do mathematics is that, if there is ever trouble, then we know where the trouble must lie.

Fortunately for us, everything in this course has been worked out in advance. We will run into no logical problems and no paradoxes in our work. In other words, mathematics is a subject that works.

It should be stressed that basic mathematics is *known* to be—indeed has been *proved to be*—logically consistent. [Strictly speaking, all notions of consistency in mathematics are relative to a higher-order system; you learn about these ideas in a course on formal mathematical logic. We shall not give a rigorous treatment of consistency in the present book.] The strict way in which we organize the subject is an important step in establishing this consistency. We shall say more about consistency, and also about independence, in the Appendix.

3.2   Elements of Set Theory

Beginning in this section, we will be doing mathematics in the way that it is usually done. That is, we shall define terms and we shall state and prove properties that they satisfy. In earlier chapters we were careful, but we were less mathematical. Sometimes we even had to say "This is the way we do it; don't worry." Many of the topics in Chapters 1 and 2 are really only best understood from the advanced perspective of mathematical logic. Now, and for the rest of this book, it is time to show how mathematics is done in practice.

We have already said what theorems, propositions, lemmas, and proofs are. Another formal ingredient of mathematical exposition is the "definition." A definition usually introduces a new piece of terminology or a new idea and *explains what it means in terms of ideas and terminology that have already been presented*. As you read this chapter, pause frequently to check that we are following this paradigm.

Definition 3.2.1 Let S and T be sets. We say that S is a *subset* of T, and we write $S \subset T$ or $T \supset S$, if

$$x \in S \Rightarrow x \in T.$$

We do not prove our definitions. There is *nothing to prove*. A definition introduces you to a new idea, or piece of terminology, or piece of notation.

EXAMPLE 3.2.2 Let $S = \{x \in \mathbb{N} : x > 3\}$ and $T = \{x \in \mathbb{N} : x^2 > 4\}$. Determine whether $S \subset T$ or $T \subset S$.

Solution: The key to success and clarity in handling subset questions is to *use the definition*. To see whether $S \subset T$, we must check whether $x \in S$ implies $x \in T$. Now if $x \in S$ then $x > 3$ hence $x^2 > 9$ so certainly $x^2 > 4$. Our syllogism is proved, and we conclude that $S \subset T$.

The reverse inclusion is false. For example, the number 3 is an element of T but is certainly not an element of S. We write $T \not\subset S$. ◇

EXAMPLE 3.2.3 Let $\mathbb{Z}$ denote the system of integers. Let $S = \{-2, 3\}$. Let $T = \{x \in \mathbb{Z} : x^3 - x^2 - 6x = 0\}$. Determine whether $S \subset T$ or $T \subset S$.

Solution: To see whether $S \subset T$, we must check whether $x \in S$ implies $x \in T$. Let $x \in S$. Then either $x = -2$ or $x = 3$. If $x = -2$ then $x^3 - x^2 - 6x = (-2)^3 - (-2)^2 - 6(-2) = 0$. Also, if $x = 3$ then $x^3 - x^2 - 6x = (3)^3 - (3)^2 - 6(3) = 0$. This verifies the syllogism $x \in S$ implies $x \in T$. Therefore $S \subset T$.

The reverse inclusion fails, for $0 \in T$ but $0 \notin S$. ◇

EXAMPLE 3.2.4 Let $S = \{x \in \mathbb{N} : x \geq 4\}$ and $T = \{x \in \mathbb{N} : x < 9\}$. Is it true that either $S \subset T$ or $T \subset S$?

Solution: Both inclusions are false. For $10 \in S$ but $10 \notin T$ and $2 \in T$ but $2 \notin S$. ◇

Proposition 3.2.5 *Let S and T be sets. Then $S = T$ if and only if both $S \subset T$ and $T \subset S$.*

Proof: If $S = T$ then, by definition, S and T have precisely the same elements. In particular, this means that $x \in S$ implies $x \in T$ and also $x \in T$ implies $x \in S$. That is, $S \subset T$ and $T \subset S$.

Now suppose that both $S \subset T$ and $T \subset S$. Seeking a contradiction, suppose that $S \neq T$. Then either there is some element of S that is not an element of T or there is some element of T that is not an element of S. The first eventuality contradicts $S \subset T$, and the second eventuality contradicts $T \subset S$. We conclude that $S = T$. □

Definition 3.2.6 We let $\emptyset$ denote the set that contains no elements. That is, $\forall x, x \notin \emptyset$. We call $\emptyset$ the *empty set*.

It may seem strange to consider a set with no elements. But this set arises very naturally in many mathematical contexts. For example, consider the set

$$S = \{x \in \mathbb{R} : x^2 < 0\}.$$

There are no real numbers with negative square. So there are no elements in this set. It is useful to be able to write $S = \emptyset$.

EXAMPLE 3.2.7 If S is any set, then $\emptyset \subset S$. To see this, notice that the statement "If $x \in \emptyset$ then $x \in S$" *must* be true because the hypothesis $x \in \emptyset$ is false. [Check the truth table for "if-then" statements.] This verifies that $\emptyset \subset S$.   ◇

EXAMPLE 3.2.8 Let $S = \{x \in \mathbb{N} : x + 2 \geq 19 \text{ and } x < 3\}$. Then S is a sensible set. There are no internal contradictions in its definition. But $S = \emptyset$. There are no elements in S.   ◇

Definition 3.2.9 Let S and T be sets. We say that x is an element of $S \cap T$ if both $x \in S$ *and* $x \in T$. It is useful and enlightening to write

$$x \in S \cap T \Longleftrightarrow x \in S \wedge x \in T.$$

This relates our new set-theoretic idea to basic concepts of logic that we learned in Chapter 1.

We say that x is an element of $S \cup T$ if either $x \in S$ *or* $x \in T$. Again, it is useful to relate the new idea to basic logic by writing

$$x \in S \cup T \Longleftrightarrow x \in S \vee x \in T.$$

We call $S \cap T$ the *intersection* of the sets S and T. We call $S \cup T$ the *union* of the sets S and T.

EXAMPLE 3.2.10 Let $S = \{x \in \mathbb{N} : 2 < x < 9\}$ and $T = \{x \in \mathbb{N} : 5 \leq x < 14\}$. Then $S \cap T = \{x \in \mathbb{N} : 5 \leq x < 9\}$, for these are the points common to both sets. And $S \cup T = \{x \in \mathbb{N} : 2 < x < 14\}$, for these are the points that are either in S or in T or in both. Draw a diagram on a real line to help you understand this example.   ◇

Remark 3.2.11 Observe that the use of "or" in the definition of set union justifies our decision to use the "inclusive 'or' " rather than the "exclusive 'or'" in mathematics.

EXAMPLE 3.2.12 Let $S = \{x \in \mathbb{N} : 1 \leq x \leq 5\}$ and $T = \{x \in \mathbb{N} : 8 < x \leq 12\}$. Then $S \cap T = \emptyset$, for the sets S and T have no elements in common. On the other hand, $S \cup T = \{x \in \mathbb{N} : 1 \leq x \leq 5 \text{ or } 8 < x \leq 12\}$. Draw a diagram on a real line to help you understand this example.   ◇

We may consider the intersection of more than two sets. For example, $x \in S \cap T \cap U$ means that (simultaneously) $x \in S$, $x \in T$, and $x \in U$. In fact one can consider the intersection of any number of sets—even an infinite number.

EXAMPLE 3.2.13 Let $S = \{1, 2, 3\}$, $T = \{2, 3, 4\}$, and $U = \{3, 4, 1\}$. Then

$$S \cap T = \{2, 3\}, \quad T \cap U = \{3, 4\}, \quad U \cap S = \{1, 3\}, \quad S \cap T \cap U = \{3\}.$$

◇

EXAMPLE 3.2.14 Let $S = \{1, 2\}$, $T = \{2, 3\}$, and $U = \{3, 1\}$. Then $S \cap T \cap U = \emptyset$. In other words, the three sets have no elements in common. But $S \cap T = \{2\} \neq \emptyset$, $T \cap U = \{3\} \neq \emptyset$, and $U \cap S = \{1\} \neq \emptyset$. It is common to say that S, T, and U are *disjoint sets*.

But S, T, and U are *not pairwise disjoint*. The phrase "pairwise disjoint" means that no two of the sets have any elements in common. For instance, $A = \{1, 2\}$, $B = \{3, 4\}$, and $C = \{5, 6\}$ *are* pairwise disjoint. ◇

Definition 3.2.15 Let S and T be sets. We say that $x \in S \setminus T$ if both $x \in S$ and $x \notin T$. We may write this logically as

$$x \in S \setminus T \Longleftrightarrow x \in S \wedge x \notin T.$$

We call $S \setminus T$ the *set-theoretic difference* of S and T.

EXAMPLE 3.2.16 Let $S = \{x \in \mathbb{N} : 2 < x < 7\}$ and $T = \{x \in \mathbb{N} : 5 \leq x < 10\}$. Then $S \setminus T = \{x \in \mathbb{N} : 2 < x < 5\}$ and $T \setminus S = \{x \in \mathbb{N} : 7 \leq x < 10\}$. ◇

Definition 3.2.17 Suppose that we are studying subsets of a fixed set X. If $S \subset X$, then we use the symbol ^cS to denote $X \setminus S$. In this context, we sometimes refer to X as the *universal set*. We call ^cS the *complement* of S (in X). We may write

$$x \in {}^cS \Longleftrightarrow x \in X \wedge x \notin S.$$

EXAMPLE 3.2.18 Let $\mathbb{N}$ be the universal set. Let $S = \{x \in \mathbb{N} : 3 < x \leq 20\}$. Then

$$^cS = \{x \in \mathbb{N} : 1 \leq x \leq 3\} \cup \{x \in \mathbb{N} : 20 < x\}.$$

◇

Proposition 3.2.19 *Let X be the universal set and $S \subset X$, $T \subset X$. Then*
(a) $^c(S \cup T) = {}^cS \cap {}^cT$;
(b) $^c(S \cap T) = {}^cS \cup {}^cT$.

Proof: We shall present this proof in detail since it is a good exercise in understanding both our definitions and our method of proof (and also a good exercise with logic).

We begin with the proof of (a). It is often best to treat the proof of the equality of two sets as two separate proofs of containment. [This is why Proposition 3.2.5 is important.] That is what we now do.

Let $x \in {}^c(S \cup T)$. Then, by definition, $x \notin (S \cup T)$. Thus x is neither an element of S nor an element of T. So both $x \in {}^cS$ and $x \in {}^cT$. Hence $x \in {}^cS \cap {}^cT$. We conclude that $^c(S \cup T) \subset {}^cS \cap {}^cT$. Conversely, if $x \subset {}^cS \cap {}^cT$, then $x \notin S$ and $x \notin T$. Therefore $x \notin (S \cup T)$. As a result, $x \in {}^c(S \cup T)$. Thus $^cS \cap {}^cT \subset {}^c(S \cup T)$. Summarizing, we have $^c(S \cup T) = {}^cS \cap {}^cT$.

The proof of part (b) is similar, but we include it for practice. Let $x \in {}^c(S \cap T)$. Then, by definition, $x \notin (S \cap T)$. Thus x is not both an element of S and an element of T. So either $x \in {}^cS$ or $x \in {}^cT$. Hence $x \in {}^cS \cup {}^cT$. We conclude that ${}^c(S \cap T) \subset {}^cS \cup {}^cT$. Conversely, if $x \in {}^cS \cup {}^cT$, then either $x \notin S$ or $x \notin T$. Therefore $x \notin (S \cap T)$. As a result, $x \in {}^c(S \cap T)$. Thus ${}^cS \cup {}^cT \subset {}^c(S \cap T)$. Summarizing, we have ${}^c(S \cap T) = {}^cS \cup {}^cT$. □

The two formulas in the last proposition are often referred to as de Morgan's Laws. Compare them with de Morgan's laws for $\vee$ and $\wedge$ in Section 1.3 and also with the de Morgan's Laws that we discussed in Section 1.7.

3.3 Venn Diagrams

We sometimes use a *Venn diagram* to aid our understanding of set-theoretic relationships. In a Venn diagram, a set is represented as a region in the plane (for convenience, we use rectangles). The intersection $A \cap B$ of two sets A and B is the region common to the two domains (we have shaded that region with dots in Figure 3.1):

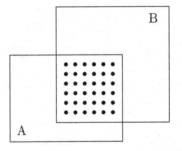

FIGURE 3.1

Now let A, B, and C be three sets. The Venn diagram in Figure 3.2 makes it easy to see that $A \cap (B \cup C) = (A \cap B) \cup (A \cap C)$.

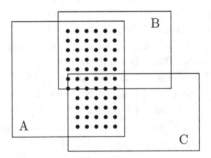

FIGURE 3.2

The Venn diagram in Figure 3.3 illustrates the fact that

$$A \setminus (B \cup C) = (A \setminus B) \cap (A \setminus C)$$

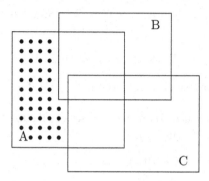

FIGURE 3.3

A Venn diagram is not a proper substitute for a rigorous mathematical proof. However, it can go a long way toward guiding our intuition.

3.4 Further Ideas in Elementary Set Theory

Now we learn some new ways to combine sets.

Definition 3.4.1 Let S and T be sets. We define $S \times T$ to be the set of all ordered pairs (s, t) such that $s \in S$ and $t \in T$. The set $S \times T$ is called the *set-theoretic product* (or sometimes just the *product*) of S and T.

It is worth pausing a moment to consider this last definition. Strictly speaking, it is not entirely satisfactory because we have not defined "ordered pair." Any attempt to do so using phrases like "first element" and "second element" seems to lead to more questions than answers. A rigorous and elegant way to define the ordered pair (a, b) is that it is equal to the set $\{\{a\}, \{a, b\}\}$. This definition exhibits directly that the ordered pair contains both the elements a and b and that the element a is distinguished.

EXAMPLE 3.4.2 Let $S = \{1, 2, 3\}$ and $T = \{a, b\}$. Then

$$S \times T = \{(1, a), (1, b), (2, a), (2, b), (3, a), (3, b)\}.$$

◇

It is no coincidence that, in the last example, the set S has 3 elements, the set T has 2 elements, and the set $S \times T$ has $3 \times 2 = 6$ elements. In fact one can prove that if S has k elements and T has ℓ elements, then $S \times T$ has $k \cdot \ell$ elements. Exercise 3.20 asks you to prove this assertion by mathematical induction on k.

Notice that $S \times T$ is a different set from $T \times S$. With S and T as in the last example,

$$T \times S = \big\{(a,1),(b,1),(a,2),(b,2),(a,3),(b,3)\big\}.$$

The phrase "ordered pair" means that the pair $(a,1)$, for example, is distinct from the pair $(1,a)$.

If S is a set then the *power set* of S is the set of all subsets of S. We denote the power set by $\mathcal{P}(S)$. We may write

$$X \in \mathcal{P}(S) \Longleftrightarrow X \subset S.$$

EXAMPLE 3.4.3 Let $S = \{1,2,3\}$. Then

$$\mathcal{P}(S) = \big\{\{1\},\{2\},\{3\},\{1,2\},\{2,3\},\{1,3\},\{1,2,3\},\emptyset\big\}.$$

◇

If the concept of power set is new to you, then you might have been surprised to see $\{1,2,3\}$ and $\emptyset$ as elements of the power set. But they are both subsets of S, and they must be listed.

Proposition 3.4.4 Let $S = \{s_1,\ldots,s_j\}$ be a set. Then $\mathcal{P}(S)$ has 2^j elements.

Proof: We prove the assertion by mathematical induction on j.

(c) The assertion $P(j)$ is this:

The power set of a set with j elements has 2^j elements.

(b) $P(1)$ is true. In this case, $S = \{s_1\}$ and $\mathcal{P}(S) = \{\{s_1\},\emptyset\}$. Notice that S has $j=1$ element and $\mathcal{P}(S)$ has $2^j = 2$ elements.

(c) $P(j) \Rightarrow P(j+1)$. Assume that any set of the form $S = \{s_1,\ldots,s_j\}$ has power set with 2^j elements. Now let $T = \{t_1,\ldots,t_j,t_{j+1}\}$. Consider the subset $T' = \{t_1,\ldots,t_j\}$ of T. Then $\mathcal{P}(T)$ certainly contains $\mathcal{P}(T')$ (that is, every subset of T' is also a subset of T). But it also contains each of the sets that is obtained by adjoining the element t_{j+1} to a subset of T'. Thus the total number of subsets of T is

$$2^j + 2^j = 2^{j+1}.$$

Notice that we have indeed counted all subsets of T, since any subset either contains t_{j+1} or it does not.

Thus, assuming the validity of our assertion for j, we have proved its validity for $j + 1$. That completes our mathematical induction and the proof of the proposition. $\square$

We have seen that the operation of set-theoretic product corresponds to the arithmetic product of natural numbers. And now we have seen that the operation of taking the power set corresponds to exponentiation. In Section 4.3 we shall use the concept of function to unify all of these ideas.

3.5 Indexing and Extended Set Operations

Frequently we wish to manipulate infinitely many sets. Perhaps we will take their intersection or union. We require suitable notation to perform these operations.

If $S_1, S_2, \ldots$ are sets, then we define

$$\bigcup_{j=1}^{\infty} S_j \equiv \left\{x : \exists j \text{ such that } x \in S_j\right\}.$$

Similarly, we define

$$\bigcap_{j=1}^{\infty} S_j \equiv \left\{x : \forall j, x \in S_j\right\}.$$

Notice that we employ the common mathematical notation $\equiv$ to mean "is defined to be." Other texts use the notation $\overset{\text{def}}{=}$ or $=:$ or $\doteq$.

EXAMPLE 3.5.1 Let $\mathbb{Q}$ be the rational numbers and let $S_j = \{x \in \mathbb{Q} : 0 < x < 1 + 1/j\}$, $j = 1, 2, \ldots$. Let us describe $\cup_{j=1}^{\infty} S_j$ and $\cap_{j=1}^{\infty} S_j$.

Notice that $S_1 \supset S_j$ for every j, hence $\cup_{j=1}^{\infty} S_j = S_1 = \{x \in \mathbb{Q} : 0 < x < 2\}$.

Next, notice that, if $x \in \mathbb{Q}$ and $x > 1$, then if we select $j > 1/(x-1)$ then $x \notin S_j$. It follows that $x \notin \cap_{j=1}^{\infty} S_j$. On the other hand, $\{x \in \mathbb{Q} : 0 < x \leq 1\} \subset S_j$ for every j. It follows that $\cap_{j=1}^{\infty} S_j = \{x \in \mathbb{Q} : 0 < x \leq 1\}$. $\diamond$

EXAMPLE 3.5.2 It is entirely possible for *nested*, non-empty sets to have empty intersection. Let $S_j = \{x \in \mathbb{Q} : 0 < x < 1/j\}$. Certainly each S_j is non-empty, for it contains the point $1/(2j)$. In fact each S_j has infinitely many elements. Next, $S_1 \supset S_2 \supset \cdots$. Finally, for any positive integer M,

$$\bigcap_{j=1}^{M} S_j = S_M \neq \emptyset.$$

However,

$$\bigcap_{j=1}^{\infty} S_j = \emptyset.$$

To verify this last assertion, notice that, if $x > 0$ and $j > 1/x$, then $x \notin S_j$, hence $x \notin \cap_{j=1}^{\infty} S_j$. However, if $x \leq 0$, then x is not an element of any S_j. As a result, no x lies in the intersection. The intersection is empty. ◇

In the examples given thus far, the "index set" has been the natural numbers. That is, we let the index j range over $\{1, 2, \dots\}$. It is frequently useful to use a larger index set, such as the real numbers or some unspecified index set. Usually we specify an index set with the letter A and we denote a specific index by $\alpha \in A$.

EXAMPLE 3.5.3 For each real number α we let $S_\alpha = \{x \in \mathbb{R} : \alpha \leq x < \alpha + 1\}$. Thus each S_α is an "interval" of real numbers, and we may speak of

$$\bigcup_{\alpha \in A} S_\alpha \equiv \{x : \exists \alpha, x \in S_\alpha\}$$

and

$$\bigcap_{\alpha \in A} S_\alpha \equiv \{x : \forall \alpha, x \in S_\alpha\}.$$

For the sets S_α that we have specified,

$$\bigcap_{\alpha \in A} S_\alpha = \emptyset.$$

This is so because, if $x \in \mathbb{R}$, then $x \notin S_{x+1}$, hence certainly $x \notin \cap_\alpha S_\alpha$.

On the other hand,

$$\bigcup_{\alpha \in A} S_\alpha = \mathbb{R}$$

since every real x lies in $S_{x-1/2}$. ◇

Proposition 3.5.4 *Fix a universal set X. Let A be an index set and, for each $\alpha \in A$, let S_α be a subset of X. Then*

(a) $\displaystyle {}^c\left(\bigcap_{\alpha \in A} S_\alpha\right) = \bigcup_{\alpha \in A} {}^c S_\alpha;$

(b) $\displaystyle {}^c\left(\bigcup_{\alpha \in A} S_\alpha\right) = \bigcap_{\alpha \in A} {}^c S_\alpha.$

Proof: The proof is similar to that of Proposition 3.2.19 of this chapter. We leave the details to Exercise 3.4 at the end of the chapter. □

Further properties of intersection and union over arbitrary index sets are explored in the exercises. These are some of the most important exercises in the chapter.

Exercises

1. Let $S = \{1,2,3,4,5\}$, $T = \{3,4,5,7,8,9\}$, $U = \{1,2,3,4,9\}$, $V = \{2,4,6,8\}$. Calculate each of the following:

   (a) $S \cap U$
   (b) $(S \cap T) \cup U$
   (c) $(S \cup U) \cap V$
   (d) $(S \cup V) \setminus U$
   (e) $(U \cup V \cup T) \setminus S$
   (f) $(S \cup V) \setminus (T \cap U)$
   (g) $(S \times V) \setminus (T \times U)$
   (h) $(V \setminus T) \times (U \setminus S)$

2. Let S be any set and let $T = \emptyset$. What can you say about $S \times T$?

3. Prove the following formulas for arbitrary sets S, T, U, and V. [**Hint:** You may find Venn diagrams useful to guide your thinking, but a Venn diagram is *not* a proof.]

   (a) $S \cap (T \cup U) = (S \cap T) \cup (S \cap U)$
   (b) $S \cup (T \cap U) = (S \cup T) \cap (S \cup U)$
   (c) $S \cap {}^cT = S \setminus T$
   (d) $(S \setminus T) \cup (T \setminus S) = (S \cup T) \setminus (S \cap T)$
   (e) $S \setminus (T \cup U) = (S \setminus T) \cap (S \setminus U)$
   (f) $S \setminus (T \cap U) = (S \setminus T) \cup (S \setminus U)$
   (g) $(S \setminus T) \times (U \setminus V) = (S \times U) \setminus [(S \times V) \cup (T \times U)]$
   (h) $(S \cup T) \times V = (S \times V) \cup (T \times V)$

4. Let $S_\alpha \subset X$ be sets indexed over an arbitrary index set A, $\alpha \in A$. Prove each of the following identities:

   (a) ${}^c\left(\cap_{\alpha \in A} S_\alpha\right) = \cup_{\alpha \in A} {}^c S_\alpha$
   (b) ${}^c\left(\cup_{\alpha \in A} S_\alpha\right) = \cap_{\alpha \in A} {}^c S_\alpha$
   (c) $T \cap \left(\cup_{\alpha \in A} S_\alpha\right) = \cup_{\alpha \in A} (T \cap S_\alpha)$
   (d) $T \cup \left(\cap_{\alpha \in A} S_\alpha\right) = \cap_{\alpha \in A} (T \cup S_\alpha)$

5. Draw Venn diagrams to illustrate parts (a)–(f) of Exercise 3 above.

6. Suppose that $A \subset B \subset C$. What is $A \setminus B$? What is $A \setminus C$? What is $A \cup B$?

In Exercises 3.7–3.9, let $\mathbb{N}$ denote the natural numbers, $\mathbb{Z}$ the integers, $\mathbb{Q}$ the rational numbers, and $\mathbb{R}$ the real numbers.

7. Describe the set $\mathbb{Q} \setminus \mathbb{Z}$ in words. Describe $\mathbb{R} \setminus \mathbb{Q}$.

8. Describe $\mathbb{Q} \times \mathbb{R}$ in words. Describe $\mathbb{Q} \times \mathbb{Z}$.

9. Describe $(\mathbb{Q} \times \mathbb{R}) \setminus (\mathbb{Z} \times \mathbb{Q})$ in words.

10. Give an explicit description of the power set of $S = \{a, b, 1, 2\}$ (that is, write out all the elements).

11. Let the set S have k elements. Give a direct proof (different from the one in the text) of the assertion that the number of elements of the power set of S is 2^k. That is, devise an explicit scheme for counting the subsets.

12. Calculate the power set of the power set of $T = \{1, 2\}$.

13. TRUE or FALSE: If $S_1, S_2, \ldots$ are sets of integers and if $\cup_{j=1}^{\infty} S_j = \mathbb{Z}$, then one of the sets S_j must have infinitely many elements. Give a proof of your answer.

14. TRUE or FALSE: If $S_1, S_2, \ldots$ are sets of real numbers and if $\cup_{j=1}^{\infty} S_j = \mathbb{R}$, then one of the sets S_j must have infinitely many elements. Give a proof of your answer.

15. Prove that $S \subset T$ if and only if $\mathcal{P}(S) \subset \mathcal{P}(T)$.

16. Prove that $S = T$ if and only if $\mathcal{P}(S) = \mathcal{P}(T)$.

17. Prove that if $A \subset B$ and $B \subset C$ then $A \subset C$.

18. Let $S = \{a, b, c, d\}$, $T = \{1, 2, 3\}$, and $U = \{b, 2\}$. Which of the following statements is true?

  (a) $\{a\} \in S$
  (b) $a \in S$
  (c) $\{a, c\} \subset S$
  (d) $\emptyset \in S$
  (e) $\{a\} \in \mathcal{P}(S)$
  (f) $\{\{a\}, \{a, b\}\} \subset \mathcal{P}(S)$
  (g) $\{a, c, 2, 3\} \subset S \cup T$
  (h) $U \subset S \cup T$
  (i) $b \in S \cap U$
  (j) $\{b\} \subset S \cap U$

      (k) $\{1,3\} \in T$

      (l) $\{1,3\} \subset T$

    (m) $\{1,3\} \in \mathcal{P}(T)$

     (n) $\emptyset \in \mathcal{P}(S)$

     (o) $\{\emptyset\} \in \mathcal{P}(S)$

     (p) $\emptyset \subset \mathcal{P}(S)$

     (q) $\{\emptyset\} \subset \mathcal{P}(S)$

19. Write out the power set of each set:

     (a) $\{1, \emptyset, \{a,b\}\}$

     (b) $\{\bullet, \triangle, \partial\}$

     (c) $\left\{\emptyset, \{\emptyset\}, \{\{\emptyset\}\}\right\}$

20. Prove by mathematical induction on k that, if the set S has k elements and the set T has ℓ elements, then $S \times T$ has $k \cdot \ell$ elements.

21. Prove or disprove each of the following statements:

     (a) $\mathcal{P}(A \cup B) = \mathcal{P}(A) \cup \mathcal{P}(B)$

     (b) $\mathcal{P}(A \cap B) = \mathcal{P}(A) \cap \mathcal{P}(B)$

     (c) $\mathcal{P}(A \setminus B) = \mathcal{P}(A) \setminus \mathcal{P}(B)$

22. If $S \subset T_\alpha$ for every $\alpha \in A$ then prove that

$$S \subset \cap_{\alpha \in A} T_\alpha$$

23. If $S \supset T_\alpha$ for every $\alpha \in A$ then prove that

$$S \supset \cup_{\alpha \in A} T_\alpha$$

24. Let $\{A_j\}$ be sets. We say that the A_j are *disjoint* if $\cap_j A_j = \emptyset$. On the other hand, the A_j are *pairwise disjoint* if $A_j \cap A_k = \emptyset$ whenever $j \neq k$. Show that these two concepts are different for four sets.

25. Let S, T, U be finite sets. Verify that the sets $(S \times T) \times U$ and $S \times (T \times U)$ have the same number of elements.

26. Let S be an infinite set and suppose that $T \subset S$. Prove that either T is infinite or $S \setminus T$ is infinite.

27. Suppose that S is a set and that $S \times T$ is finite for every choice of finite set T. Prove that then S must be finite.

28. Suppose that S is a set and that $S \cup T$ is finite for every choice of finite set T. Prove that then S must be finite.

* **29.** Let $\mathcal{P}$ be the power set of $I = \{1, 2, 3, 4, 5\}$. Let $\mathcal{S}$ be a randomly chosen subset of $\mathcal{P}$. What is the probability that $\mathcal{S}$ is the power set of some subset of I?

4

Relations and Functions

4.1 Relations

Let S and T be sets. A *relation* on S and T is a subset of $S \times T$. If $\mathcal{R}$ is a relation, then we write either $(s, t) \in \mathcal{R}$ or sometimes $s \, \mathcal{R} \, t$ to indicate that s is related to t or that (s, t) is an element of the relation. We will also write $s \sim t$ when the relation being discussed is understood.

EXAMPLE 4.1.1 Let $S = \mathbb{N}$, the natural numbers; and let $T = \mathbb{R}$, the real numbers. Define a relation $\mathcal{R}$ on S and T by $(s, t) \in \mathcal{R}$ if $s < \sqrt{t} < s + 1$. For instance, $(2, 5) \in \mathcal{R}$ because $\sqrt{5}$ lies between 2 and 3. Also $(4, 17) \in \mathcal{R}$ because $\sqrt{17}$ lies between 4 and 5. However, $(5, 10)$ does not lie in $\mathcal{R}$. Also $(3, \pi)$ does not lie in $\mathcal{R}$. ◇

The *domain* of a relation $\mathcal{R}$ is the set of $s \in S$ such that there exists a $t \in T$ with $(s, t) \in \mathcal{R}$. The *image* of the relation is the set of $t \in T$ such that there exists an $s \in S$ with $(s, t) \in \mathcal{R}$. It is sometimes convenient to refer to the entire set T as the *range* of the relation $\mathcal{R}$. So we see that the image and the range are distinct. Some sources use the word "codomain" rather than "range."

EXAMPLE 4.1.2 Let $S = \mathbb{N}$ and $T = \mathbb{N}$. Define a relation $\mathcal{R}$ on S and T by the condition $(s, t) \in \mathcal{R}$ if $s^2 < t$. Observe that, for any element $s \in \mathbb{N} = S$, the number $t = s^2 + 1$ satisfies $s^2 < t$. Therefore every $s \in S = \mathbb{N}$ is in the domain of the relation.

Now let us think about the image. The number $1 \in \mathbb{N} = T$ cannot be in the image since there is no element $s \in S = \mathbb{N}$ such that $s^2 < 1$. However, any element $t \in T$ that exceeds 1 satisfies $1^2 < t$. So $(1, t) \in \mathcal{R}$. Thus the image of $\mathcal{R}$ is the set $\{t \in \mathbb{N} : t \geq 2\}$. The range of the relation can be taken to be $\mathbb{N}$ itself. ◇

EXAMPLE 4.1.3 Let $S = \mathbb{N}$ and $T = \mathbb{N}$. Define a relation $\mathcal{R}$ on S and T by the condition $(s, t) \in \mathcal{R}$ if $s^2 + t^2$ is itself a perfect square. Then, for instance, $(3, 4) \in \mathcal{R}$, $(4, 3) \in \mathcal{R}$, $(12, 5) \in \mathcal{R}$, and $(5, 12) \in \mathcal{R}$. The number 1 is not in the domain of $\mathcal{R}$ since there is no natural number t such that $1^2 + t^2$ is a perfect square (if there were, this would mean that there are two perfect

DOI: 10.1201/9781003214564-4

squares that differ by 1, and that is not the case). The number 2 is not in the domain of $\mathcal{R}$ for a similar reason. Likewise, 1 and 2 are not in the image of $\mathcal{R}$.

In fact, both the domain and image of $\mathcal{R}$ have infinitely many elements. This assertion will be explored in Exercise 4.56.   ◇

Many interesting relations arise for which S and T are the same set. Say that $S = T = A$. Then a relation on S and T is called simply a relation on A.

EXAMPLE 4.1.4 Let $\mathbb{Z}$ be the integers. Let us define a relation $\mathcal{R}$ on $\mathbb{Z}$ by the condition $(s,t) \in \mathcal{R}$ if $s - t$ is divisible by 2. It is easy to see that both the domain and the image of this relation is $\mathbb{Z}$ itself. It is also worth noting that, if n is any integer, then the set of all elements related to n is either (i) the set of all even integers (if n is even) or (ii) the set of all odd integers (if n is odd).   ◇

Notice that the last relation created a division of the domain (=image) into two disjoint sets: the even integers and the odd integers. This was a special instance of an important type of relation that we now define.

Definition 4.1.5 Let $\mathcal{R}$ be a relation on a set A. We say that $\mathcal{R}$ is an *equivalence relation* if the following properties hold:

$\mathcal{R}$ **is reflexive:** If $x \in A$, then $(x,x) \in \mathcal{R}$;

$\mathcal{R}$ **is symmetric:** If $(x,y) \in \mathcal{R}$, then $(y,x) \in \mathcal{R}$;

$\mathcal{R}$ **is transitive:** If $(x,y) \in \mathcal{R}$ and $(y,z) \in \mathcal{R}$, then $(x,z) \in \mathcal{R}$.

Check for yourself that the relation described in Example 4.1.4 is in fact an equivalence relation. The most important property of equivalence relations is that which we indicated just before the definition and which we now enunciate formally:

Proposition 4.1.6 *Let $\mathcal{R}$ be an equivalence relation on a set A. If $x \in A$, then define*

$$E_x \equiv \{y \in A : (x,y) \in \mathcal{R}\}.$$

We call the sets E_x the equivalence classes induced by the relation $\mathcal{R}$. If s and t are any two elements of A, then either $E_s \cap E_t = \emptyset$ or $E_s = E_t$.

In summary, the set A is the pairwise disjoint union of the equivalence classes induced by the equivalence relation $\mathcal{R}$.

Before we prove this proposition, let us discuss for a moment what it means. Clearly every element $a \in A$ is contained in some equivalence class, for a is contained in E_a itself. The proposition tells us that the set A is in fact the pairwise disjoint union of these equivalence classes. We say that the equivalence classes *partition* the set A.

For instance, in Example 4.1.4, the equivalence relation gives rise to two equivalence classes: the even integers $\mathcal{E}$ and the odd integers $\mathcal{O}$. Of course $\mathbb{Z} = \mathcal{E} \cup \mathcal{O}$ and $\mathcal{E} \cap \mathcal{O} = \emptyset$. We say that the equivalence relation *partitions* the universal set $\mathbb{Z}$ into two equivalence classes.

Notice that, in Example 4.1.4, if we pick any element $x \in \mathcal{E}$, then $E_x = \mathcal{E}$. Likewise, if we pick any element $y \in \mathcal{O}$, then $E_y = \mathcal{O}$.

Proof of the Proposition: Let $s, t \in A$ and suppose that $E_s \cap E_t \neq \emptyset$. It is our job to prove that $E_s = E_t$ (think for a moment about the truth table for "or" so that you understand that we are doing the right thing).

Since $E_s \cap E_t \neq \emptyset$, there is an element $x \in E_s \cap E_t$. Then $x \in E_s$. Therefore, by definition, $(s, x) \in \mathcal{R}$. Likewise, $x \in E_t$. Thus $(t, x) \in \mathcal{R}$. By symmetry, it follows that $(x, t) \in \mathcal{R}$. Now transitivity tells us that, since $(s, x) \in \mathcal{R}$ and $(x, t) \in \mathcal{R}$, then $(s, t) \in \mathcal{R}$.

If y is any element of E_t, then $(t, y) \in \mathcal{R}$. Transitivity now implies that since $(s, t) \in \mathcal{R}$ and $(t, y) \in \mathcal{R}$, then $(s, y) \in \mathcal{R}$. Thus $y \in E_s$. We have shown that every element of E_t is an element of E_s. Thus $E_t \subset E_s$.

Reversing the roles of s and t, we find that $E_s \subset E_t$. It follows that $E_s = E_t$. This is what we wished to prove. □

EXAMPLE 4.1.7 Let A be the set of all people in the United States. If $x, y \in A$, then let us say that $(x, y) \in \mathcal{R}$ if x and y have the same surname (i.e., last name). Then $\mathcal{R}$ is an equivalence relation:

(i) $\mathcal{R}$ is reflexive since any person x has the same surname as his/her self.
(ii) $\mathcal{R}$ is symmetric since if x has the same surname as y, then y has the same surname as x.
(iii) $\mathcal{R}$ is transitive since if x has the same surname as y and y has the same surname as z, then x has the same surname as z.

Thus $\mathcal{R}$ is an equivalence relation. The equivalence classes are all those people with surname Smith, all those people with surname Herkimer, and so forth. ◇

EXAMPLE 4.1.8 Let S be the set of all residents of the United States. If $x, y \in S$, then let us say that x is related to y (that is, $x \sim y$) if x and y have at least one biological parent in common. It is easy to see that this relation is reflexive and symmetric. It is *not* transitive, as children of divorced parents know too well. What this tells us (mathematically) is that the proliferation of divorce in our society does *not* lead to well-defined families. ◇

EXAMPLE 4.1.9 Let S be the set of all residents of the United States. If $x, y \in S$, then let us say that x is related to y (that is, $x \sim y$) if x and y have *both* biological parents in common. It is easy to see that this relation is reflexive and symmetric. It is also transitive, since if A has the same Mom and Dad as B and B has the same Mom and Dad as C, then A, B, C are siblings and A has the same Mom and Dad as C. Contrast this situation with that in the last example.

What this tells us (mathematically) is that traditional families are defined by an equivalence relation.   ◇

EXAMPLE 4.1.10 Let S be the set of integers and say that $x \sim y$ if $x \leq y$. This relation is clearly reflexive. It is *not* symmetric, as $3 \leq 5$ but $5 \not\leq 3$. You may check that it is transitive. But the failure of symmetry tells us that this is not an equivalence relation.   ◇

EXAMPLE 4.1.11 Let f be a function with domain the real numbers and range the real numbers. We say that two numbers $a, b \in \mathbb{R}$ are related if $f(a) = f(b)$. This relation is clearly reflexive and symmetric. Also, if $f(a) = f(b)$ and $f(b) = f(c)$, then $f(a) = f(c)$. So the relation is also transitive, and it is therefore an equivalence relation. The equivalence classes are called *inverse images of points in the range*. For example, the set of all x such that $f(x) = 5$ is an equivalence class. It is the inverse image of 5.   ◇

EXAMPLE 4.1.12 Let

$$S = \{1, 2, 3, 4, 5, 6, 7, 8, 9, 10\}.$$

Define
$$E_1 = \{1, 4, 7, 10\} \quad E_2 = \{2, 5, 8\} \quad E_3 = \{3, 6, 9\}.$$

Then the sets E_1, E_2, E_3 are pairwise disjoint, and their union is S. So these could be the equivalence classes for an equivalence relation, and in fact they are. What is that relation?

Say that $a \sim b$ if $b - a$ is divisible by 3. Check for yourself that this relation is reflexive, symmetric, and transitive. And verify that the equivalence class of 1 is E_1, the equivalence class of 2 is E_2, and the equivalence class of 3 is E_3.   ◇

This last example is an instance of a general phenomenon. If a set S is partitioned into subsets (pairwise disjoint sets whose union is S), then those subsets will be the equivalence classes for an equivalence relation.

4.2   Order Relations

In this section we discuss the concept of ordering a set. There are many different types of orderings; we shall concentrate on just a few of these.

Definition 4.2.1 Let S be a set and $\mathcal{R}$ a relation on S. We call $\mathcal{R}$ a *partial ordering* on S if it satisfies the following properties:

(a) For all $x \in S$, $(x, x) \in \mathcal{R}$;

(b) If $x, y \in S$ and both $(x, y) \in \mathcal{R}$ and $(y, x) \in \mathcal{R}$, then $x = y$;

(c) If $x, y, z \in S$ and both $(x, y) \in \mathcal{R}$ and $(y, z) \in \mathcal{R}$, then $(x, z) \in \mathcal{R}$.

Not surprisingly, we refer to property **(a)** as *reflexivity*, **(b)** as *anti-symmetry*, and **(c)** as *transitivity*.

EXAMPLE 4.2.2 Let S be the power set of $\mathbb{R}$—the set of all sets of real numbers. If $A, B \in S$, then let us say that $(A, B) \in \mathcal{R}$ if $A \subset B$, that is, if every element of A is also an element of B. Check that the three axioms for a partial ordering are satisfied by our relation $\mathcal{R}$. We usually write this relation in the binary form $A \subset B$.   ◇

A noteworthy feature of a *partial ordering* (as opposed to a total ordering, to be discussed below) is that not every two elements of the set S need be comparable. The last example illustrates this point: if $A = \{x \in \mathbb{R} : 1 < x < 4\}$ and $B = \{x \in \mathbb{R} : 2 < x < 9\}$, then both $A, B \in S$ yet neither $(A, B) \in \mathcal{R}$ nor $(B, A) \in \mathcal{R}$.

Definition 4.2.3 Let S be a set and $\mathcal{R}$ a relation on S. We call $\mathcal{R}$ a *simple ordering* on S if it satisfies the following properties:

(a) If $x, y \in S$ and both $(x, y) \in \mathcal{R}$ and $(y, x) \in \mathcal{R}$, then $x = y$;

(b) If $x, y, z \in S$ and both $(x, y) \in \mathcal{R}$ and $(y, z) \in \mathcal{R}$, then $(x, z) \in \mathcal{R}$;

(c) If $x, y \in S$ are distinct, then either $(x, y) \in \mathcal{R}$ or $(y, x) \in \mathcal{R}$.

Observe that property **(c)** distinguishes a simple ordering from a partial ordering.

As before, we refer to **(a)** and **(b)** as the properties of anti-symmetry and transitivity, respectively. We refer to property **(c)** as *strong connectivity*. A simple ordering is sometimes also called a *total ordering*.

EXAMPLE 4.2.4 Let $S = \mathbb{R}$, the real numbers. Let us say that $(x, y) \in \mathcal{R}$ if $y - x \geq 0$. It is straightforward to verify that properties **(a)**, **(b)**, **(c)** of a simple ordering hold for this relation. We usually write this relation in the binary form $x \leq y$.   ◇

Definition 4.2.5 Let us say that a set S is *strictly simply ordered* by a relation $\mathcal{R}$ if properties **(b)** and **(c)** of the last definition hold but property **(a)** is replaced by

(a′) If $x, y \in S$ and $(x, y) \in \mathcal{R}$ then $(y, x) \notin \mathcal{R}$.

EXAMPLE 4.2.6 The real numbers $S = \mathbb{R}$ are strictly simply ordered by the binary relation $<$.   ◇

Suppose that S is a set that is equipped with a strict, simple ordering $\mathcal{R}$. Let $A \subset S$. An element $a \in A$ is called *minimal* (for this ordering) in A if $(a, x) \in \mathcal{R}$ for all $x \in A$, $x \neq a$. We also sometimes call a the *least* element of A. It is clear from this definition that the minimal element is unique if it exists.

EXAMPLE 4.2.7 Let $S = \mathbb{N}$, the natural numbers. If $m, n \in \mathbb{N}$ then we say that $m\mathcal{R}n$ if $m < n$. This is a strict, simple ordering. Then $T = \{8, 4, 9, 17, 3\}$ is a subset of $\mathbb{N}$, and 3 is the minimal element of T.   ◇

Definition 4.2.8 Let us say that a strict, simple ordering $\mathcal{R}$ *well orders* a set S if each non-empty subset $A \subset X$ has a minimal element.

EXAMPLE 4.2.9 The usual ordering $<$ well orders the natural numbers. That is to say, each non-empty subset of the natural numbers $\{1, 2, 3, \dots\}$ has a minimal element. This statement is intuitively clear, but proving it quickly leads to deep and difficult questions about the foundations of mathematics. We shall treat the issue in greater detail in Section 5.1.

The ordering $<$ does *not* well order the integers $\mathbb{Z}$, nor does it well order the real numbers $\mathbb{R}$. For example $\mathcal{E}$, the even numbers, is a subset of $\mathbb{Z}$ and is also a subset of $\mathbb{R}$; but $\mathcal{E}$ certainly has no least element. The subset $S = \{x \in \mathbb{R} : 0 < x < 1\}$ (with the usual ordering on $\mathbb{R}$) has no least element.

In fact, one way to well order the integers is to construct a one-to-one correspondence between the integers and the natural numbers and then to pull the natural ordering from the natural numbers back to the integers by way of this correspondence. This gives a well ordering of the integers, but it is certainly not the standard ordering. We shall say more about this technique in Section 5.2.

It is impossible to explicitly specify a well ordering for the real numbers $\mathbb{R}$, although such a well ordering *does* exist. In fact absolutely any set can be well ordered (although it is often not at all clear how to actually perform the ordering). This matter is intimately connected with the so-called Axiom of Choice. We shall discuss that axiom in the Appendix.   ◇

We shall have more to say about orderings in Sections 4.5 and 5.2.

4.3   Functions

In more elementary mathematics courses, we define a function as follows: Let S and T be sets. A function f from S to T is a rule that assigns to each element of S a unique element of T.

This definition is problematic. The main difficulty is the use of the words "rule" and "assign." For instance, let $S = T = \mathbb{Z}$. Consider

$$f(x) = \begin{cases} x^2 & \text{if there is life as we know it on Mars} \\ 3x - 5 & \text{if there is not life as we know it on Mars.} \end{cases}$$

Is this a function? Can what we see on the right be considered a rule? Do we have to wait until we have found life on Mars before we can consider this a function?

More significantly in practice, thinking of a function as a rule is extremely limiting. The functions

$$\begin{aligned} f(x) &= x^3 - 3x + 1 \\ g(x) &= \sin x \\ h(x) &= \frac{\ln x}{x^2 + 4} \end{aligned}$$

are inarguably given by rules. But open up your newspaper and look on the financial page at the graph of the Gross National Product. This is certainly the graph of a function, but what "rule" describes it?

It is best in advanced mathematics to have a way to think about functions that avoids subjective words like "rule" and "assign." This is the motivation for our next definition.

Definition 4.3.1 Let S and T be sets. A *function f* from S to T is a relation on S and T such that

(i) Every $s \in S$ is in the domain of f;

(ii) If $(s, t) \in f$ and $(s, u) \in f$ then $t = u$.

Of course we refer to S and T as the domain and the range, respectively,[1] of f. Condition (i) mandates that each element s of S is associated to *some* element of T. Condition (ii) mandates, in a formal manner, that each element s of S is associated to *only one* member of T. Notice, however, that the definition neatly sidesteps the notions of "assign" or "rule." Now look back at our "Mars" definition and decide whether it is a function.

We shall frequently speak of the *image* of a given function f from S to T. This just means the set that is the image of f when it is thought of as a relation. It is the set of elements $t \in T$ such that there is an $s \in S$ with $(s, t) \in f$.

EXAMPLE 4.3.2 Let $S = \{1, 2, 3\}$ and $T = \{a, b, c\}$. Set

$$f = \{(1, a), (2, a), (3, b)\}.$$

---

[1] We note, once again, that some sources use the word "codomain" instead of "range."

This is a function, for it satisfies the properties set down in Definition 4.3.1. Given the way that you are accustomed to writing functions in earlier courses, you might find it helpful to view this function as

$$f(1) = a$$
$$f(2) = a$$
$$f(3) = b.$$

Notice that each element $1, 2, 3$ of the domain is "assigned" to one and only one element of the range. However, the definition of function allows the possibility that two different elements of the domain be assigned to the same range element. Observe that, for this function, the image is $\{a, b\}$ while the range is $\{a, b, c\}$.   ◇

EXAMPLE 4.3.3 Let $S = \{1, 2, 3\}$ and $T = \{a, b, c, d, e\}$. Set

$$f = \{(1, b), (2, c), (3, e)\}.$$

This is a function, for it satisfies the properties set down in Definition 4.3.1. Notice that each element of the domain S is used once and only once. However, not all elements of the range are used. According to the definition of function, this is allowed.   ◇

EXAMPLE 4.3.4 Let $S = \{1, 2, 3, 4, 5\}$ and let $T = \{a, b, c\}$. This time there are more elements in S than there are in T. Nonetheless,

$$f = \{(1, a), (2, a), (3, b), (4, b), (5, c)\}$$

is a function. It repeats values, but it definitely satisfies Definition 4.3.1.   ◇

Definition 4.3.5 Let f be a function with domain S and range T. We often write such a function as $f : S \to T$. We say that f is *one-to-one* or *injective* if, whenever $(s, t) \in f$ and $(s', t) \in f$, then $s = s'$. We sometimes refer to such a mapping as an *injection*. We also call such a map *injective* or *univalent*.

Compare this new definition with Definition 4.3.1 of function. The new condition is similar to condition (ii) for functions. But it is *not* the same. We are now mandating that no two domain elements be associated with the same range element.

EXAMPLE 4.3.6 Let $S = T = \mathbb{R}$ and let f be the set of all ordered pairs $\{(x, x^2) : x \in \mathbb{R}\}$. We may also write this function as

$$f : \mathbb{R} \to \mathbb{R}$$
$$x \mapsto x^2$$

or as $f(x) = x^2$.

It is easy to verify that f satisfies the definition of function. However, both of the ordered pairs $(-2, 4)$ and $(2, 4)$ are in f (in other words, $f(-2) = 4 = f(2)$) so that f is *not* one-to-one.   ◇

EXAMPLE 4.3.7 Let $S = T = \mathbb{R}$ and let f be the function $f(x) = x^3 + x - 5$. Then $f'(x) = 3x^2 + 1 > 0$ for every x. Therefore f is a strictly increasing function. In particular, if $s < t$, then $f(s) < f(t)$ so that $f(s) \neq f(t)$. It follows that the function f *is* one-to-one. ◇

Definition 4.3.8 Let f be a function with domain S and range T. If, for each $t \in T$ there is an $s \in S$ such that $f(s) = t$, then we say that f is *onto* or *surjective*. We sometimes refer to such a mapping as a *surjection*. Notice that a function is onto precisely when its image equals its range.

EXAMPLE 4.3.9 Let $f(x) = x^2$ be the function from Example 4.3.6. Recall from that example that $S = T = \mathbb{R}$. The point $t = -1 \in T$ has the property that there is no $s \in S$ such that $f(s) = t$. As a result, this function f is *not onto*. ◇

EXAMPLE 4.3.10 Let $S = \mathbb{R}$, $T = \{x \in \mathbb{R} : 1 \leq x < \infty\}$. Let $g : S \to T$ be given by $g(x) = x^2 + 1$. Then for each $t \in T$ the number $s = +\sqrt{t-1}$ makes sense and lies in S. Moreover, $g(s) = t$. It follows that this function g is surjective. However, g is not injective. ◇

EXAMPLE 4.3.11 Have another look at the function $f : \mathbb{R} \to \mathbb{R}$ given by $f(x) = x^2$. We have already noted, in Example 4.3.9, that this function is not onto. But if we restrict the range to $T = \{y \in \mathbb{R} : y \geq 0\}$, so that $f : \mathbb{R} \to T$, then it is easy to verify that the function is now onto. In other words, every nonnegative real number has a square root. ◇

4.4   Combining Functions

There are several elementary operations that allow us to combine functions in useful ways. In this section, and from now on, we shall (whenever possible) write our functions in the form

$$f(x) = \text{(formula)}$$

for the sake of clarity. However, we must keep in mind, and we shall frequently see, that many functions *cannot* be expressed with an elegant formula.

Definition 4.4.1 Let f and g be functions with the same domain S and the same range T. Assume that T is a set in which the indicated arithmetic operations (below) makes sense. Then we define

(a) $(f + g)(x) = f(x) + g(x)$;

(b) $(f - g)(x) = f(x) - g(x)$;

(c) $(f \cdot g)(x) = f(x) \cdot g(x)$;

(d) $(f/g)(x) = f(x)/g(x)$ provided that $g(x) \neq 0$.

Notice that, in each of **(a)**–**(d)**, we are defining a *new function*—either $f + g$ or $f - g$ or $f \cdot g$ or f/g—in terms of the component functions f and g. For practice, we shall express **(a)** in the language of ordered pairs. We ask you to do likewise with **(b)**, **(c)**, **(d)** in Exercise 4.15.

Let us consider part **(a)** in detail. Now f is a collection of ordered pairs in $S \times T$ that satisfy the conditions for a function, and so is g. The function $f + g$ is given by

$$f + g = \{(s, t + t') : (s, t) \in f, (s, t') \in g\}.$$

Expressing the other combinations of f and g is quite similar, and you should be sure to do the corresponding Exercise 4.15.

EXAMPLE 4.4.2 Let $S = T = \mathbb{R}$. Define

$$f(x) = x^3 - x \qquad \text{and} \qquad g(x) = \sin(x^2).$$

Let us calculate $f + g$, $f - g$, $f \cdot g$, f/g.
Now

$$
\begin{aligned}
(f + g)(x) &= (x^3 - x) + \sin(x^2) \\
(f - g)(x) &= (x^3 - x) - \sin(x^2) \\
(f \cdot g)(x) &= (x^3 - x) \cdot [\sin(x^2)] \\
(f/g)(x) &= (x^3 - x)/[\sin(x^2)] \quad \text{provided } x \neq \pm\sqrt{k\pi}, \\
&\qquad k \in \{0, 1, 2, \dots\}.
\end{aligned}
$$

$\diamond$

A more interesting, and more powerful, way to combine functions is through functional composition. Incidentally, in this discussion we will see the value of good mathematical notation.

Definition 4.4.3 Let $f : S \to T$ be a function, and let $g : T \to U$ be a function. Then we define, for $s \in S$, the composite function

$$(g \circ f)(s) = g(f(s)). \tag{$*$}$$

We call $g \circ f$ the *composition* of the functions g and f.

Notice in this definition that the right-hand side of $(*)$ always makes sense because of the way that we have specified the domain and range of the component functions f and g. In particular, we must have image $f \subset$ domain g in order for the composition to make sense.

EXAMPLE 4.4.4 Let $f : \mathbb{R} \to \{x \in \mathbb{R} : x \geq 0\}$ be given by $f(x) = x^4 + x^2 + 6$ and $g : \{x \in \mathbb{R} : x \geq 0\} \to \mathbb{R}$ be given by $g(x) = \sqrt{x} - 4$. Notice that f and g fit the paradigm specified in the definition of composition of functions. Then

$$
\begin{aligned}
(g \circ f)(x) &= g(f(x)) \\
&= g(x^4 + x^2 + 6) \\
&= \sqrt{x^4 + x^2 + 6} - 4.
\end{aligned}
$$

Notice that $f \circ g$ also makes sense and is given by

$$
\begin{aligned}
(f \circ g)(x) &= f(g(x)) \\
&= f(\sqrt{x} - 4) \\
&= [\sqrt{x} - 4]^4 + [\sqrt{x} - 4]^2 + 6.
\end{aligned}
$$

It is important to understand that $f \circ g$ and $g \circ f$, when both make sense, will generally be different. ◇

It is a good exercise in the ideas of this chapter to express the notion of functional composition in the language of ordered pairs. Thus let $f : S \to T$ be a function and $g : T \to U$ be a function. Then f is a subset of $S \times T$ and g is a subset of $T \times U$, both satisfying the two standard conditions for function. Now $g \circ f$ is a set of ordered pairs specified by

$$g \circ f =$$
$$\{(s, u) : s \in S, u \in U, \text{ and } \exists t \in T \text{ such that } (s, t) \in f$$
$$\text{and } (t, u) \in g\}.$$

Take a moment to verify that this equation is consistent with the definition of functional composition that we gave earlier. Further note that $g \circ f$ is a set of ordered pairs from $S \times U$.

EXAMPLE 4.4.5 Let $f : \mathbb{R} \to [-1, 1]$ be given by $f(x) = \sin x^5$ and let $g : \{x \in \mathbb{R} : x \geq 1\} \to \mathbb{R}$ be given by $\sqrt[4]{x - 1}$. We cannot consider $g \circ f$ because the range of f (namely, the set $[-1, 1]$) does not lie in the domain of g. However, $f \circ g$ *does* make sense because the range of g lies in the domain of f. And

$$(f \circ g)(x) = \sin[(x - 1)^{5/4}].$$

◇

Definition 4.4.6 Let S and T be sets. Let $f : S \to T$ and $g : T \to S$. We say that f and g are *mutually inverse* provided that both $(f \circ g)(t) = t$ for all $t \in T$ and $(g \circ f)(s) = s$ for all $s \in S$. We write $g = f^{-1}$ or $f = g^{-1}$. We refer to the functions f and g as *invertible*; we call g the *inverse* of f and f the *inverse* of g.

EXAMPLE 4.4.7 Let $f : \mathbb{R} \to \mathbb{R}$ be given by $f(x) = x^3 - 1$ and $g : \mathbb{R} \to \mathbb{R}$ be given by $g(x) = \sqrt[3]{x+1}$. Then

$$
\begin{aligned}
(f \circ g)(x) &= \left[\sqrt[3]{x+1}\right]^3 - 1 \\
&= (x+1) - 1 \\
&= x
\end{aligned}
$$

and

$$
\begin{aligned}
(g \circ f)(x) &= \sqrt[3]{(x^3 - 1) + 1} \\
&= \sqrt[3]{x^3} \\
&= x
\end{aligned}
$$

for all x. Thus $g = f^{-1}$ (or $f = g^{-1}$).   ◇

The idea of inverse function lends itself particularly well to the notation of ordered pairs. For $f : S \to T$ is inverse to $g : T \to S$ (and vice versa) provided that for every ordered pair $(s, t) \in f$ there is an ordered pair $(t, s) \in g$ and conversely.

Not every function has an inverse. For instance, let $f : S \to T$. Suppose that $f(s) = t$ and also that $f(s') = t$ with $s \neq s'$ (in other words, suppose that f is not one-to-one). If $g : T \to S$, then $g(f(s)) = g(t) = g(f(s'))$ so it cannot be that both $g(f(s)) = s$ and $g(f(s')) = s'$. In other words, f cannot have an inverse. We conclude that a function that *does* have an inverse must be one-to-one.

On the other hand, suppose that $t \in T$ has the property that there is no $s \in S$ with $f(s) = t$ (in other words, suppose that f is not onto). Then, in particular, it could not be that $f(g(t)) = t$ for any function $g : T \to S$. So f could not be invertible. We conclude that a function that *does* have an inverse must be onto.

EXAMPLE 4.4.8 Let $f : \mathbb{R} \to \{x \in \mathbb{R} : x \geq 0\}$ be given by $f(x) = x^2$. Then f is onto, but f is not one-to-one. It follows that f cannot have an inverse. And indeed it does not, for any attempt to produce an inverse function runs into the ambiguity that every positive number has two square roots.

Let $f : \{x \in \mathbb{R} : x \geq 0\} \to \mathbb{R}$ be given by $f(x) = x^2$. Then f is one-to-one but f is not onto. There certainly is a function $g : \mathbb{R} \to \{x \in \mathbb{R} : x \geq 0\}$ such that $(g \circ f)(x) = x$ for all $x \in \{x \in \mathbb{R} : x \geq 0\}$ (namely $g(x) = \sqrt{|x|}$). But there is no function $g : \mathbb{R} \to \{x \in \mathbb{R} : x \geq 0\}$ such that $(f \circ g)(x) = x$ for all x.   ◇

We have established that *if* $f : S \to T$ has an inverse, then f must be one-to-one and onto. The converse is true too, and we leave the details for you to verify. A function $f : S \to T$ that is one-to-one and onto (and therefore invertible) is sometimes called a *set-theoretic isomorphism* or a *bijection*.

EXAMPLE 4.4.9 The function $f : \mathbb{R} \to \mathbb{R}$ given by $f(x) = 3x + 5$ is one-to-one and onto (you should check this). Therefore it is invertible. To find the inverse, we consider the equation

$$f \circ f^{-1}(x) = x$$

or

$$f(f^{-1}(x)) = x.$$

We may write this out as

$$3f^{-1}(x) + 5 = x.$$

Solving for $f^{-1}(x)$ gives

$$f^{-1}(x) = \frac{x - 5}{3}.$$

Check for yourself that $f \circ f^{-1}(x) = x$ and $f^{-1} \circ f(x) = x.$ ◇

EXAMPLE 4.4.10 The function $f(x) = x^3 + 1$ is one-to-one and onto. Therefore it is invertible. To find the inverse, we consider the equation

$$f \circ f^{-1}(x) = x$$

or

$$f(f^{-1}(x)) = x.$$

We may write this out as

$$[f^{-1}(x)]^3 + 1 = x.$$

Solving for $f^{-1}(x)$ gives

$$f^{-1}(x) = \sqrt[3]{x - 1}.$$

◇

EXAMPLE 4.4.11 In this example we use some ideas from calculus.

The function $f(x) = x^3 + x$ satisfies $f'(x) = 3x^2 + 1 > 0$ for every x. Therefore f is strictly increasing. So it is one-to-one. Also $f(x) \to +\infty$ as $x \to +\infty$ and $f(x) \to -\infty$ as $x \to -\infty$. So f is onto. Therefore f is invertible. It would be quite difficult to solve the equation

$$f \circ f^{-1}(x) = x,$$

and we shall not attempt to do so. ◇

EXAMPLE 4.4.12 The function $f : \mathbb{R} \to \mathbb{R}$ that is given by $f(x) = x^3$ is a bijection. You should check the details of this assertion for yourself. The inverse of this function f is the function $g : \mathbb{R} \to \mathbb{R}$ given by $g(x) = x^{1/3}$. ◇

We leave it as an exercise for you to verify that the composition of two bijections (when the composition makes sense) is a bijection.

4.5   Cantor's Notion of Cardinality

One of the most profound ideas of modern mathematics is Georg Cantor's theory of the infinite (Georg Cantor, 1845–1918). Cantor's insight was that infinite sets can be compared by size, just as finite sets can. For instance, we think of the number 2 as *less* than the number 3; so a set with two elements is "smaller" than a set with three elements. We would like to have a similar notion of comparison for infinite sets. In this section we will present Cantor's ideas; we will also give precise definitions of the terms "finite" and "infinite."

Definition 4.5.1 Let A and B be sets. We say that A and B have the *same cardinality* if there is a function f from A to B which is both one-to-one and onto (that is, f is a bijection from A to B). We write $\operatorname{card}(A) = \operatorname{card}(B)$.

EXAMPLE 4.5.2 Let $A = \{1, 2, 3, 4, 5\}$, $B = \{\alpha, \beta, \gamma, \delta, \varepsilon\}$, and $C = \{a, b, c, d, e, f\}$. Then A and B have the same cardinality because the function

$$f = \{(1, \alpha), (2, \beta), (3, \gamma), (4, \delta), (5, \varepsilon)\}$$

is a bijection of A to B. This function is not the *only* bijection of A to B (can you find another?), but we are only required to produce one.

On the other hand, A and C do not have the same cardinality; neither do B and C. You may check these assertions by using the pigeonhole principle. ◇

Georg ·Cantor (1845–1918)

Georg Waldemar Cantor was born in Denmark. Cantor attended primary school in St. Petersburg, then in 1856 the family moved to Germany. At first they lived in Wiesbaden, then they moved to Frankfurt. After attending the Höhere Gewerbeschule in Darmstadt, from 1860 he entered the Polytechnic of Zürich in 1862. Cantor attended lectures by Weierstrass, Kummer, and Kronecker. In 1869 he presented his thesis on number theory, and received his habilitation.

At Halle the direction of Cantor's research turned away from number theory and toward analysis. This was due to Heine, who challenged Cantor to establish the uniqueness of trigonometric series. Cantor solved this problem in April, 1870. Cantor was promoted to Extraordinary Professor at Halle in 1872 and in that year began a friendship with Dedekind.

In 1873, Cantor proved the rational numbers to be countable. He also showed that the algebraic numbers are countable. In 1874, he was able to prove that the real numbers are uncountable.

Liouville established in 1851 that transcendental numbers exist. Twenty years later, in his 1874 work, Cantor showed that, in a certain sense, "almost all" numbers are transcendental.

A major paper on dimension which Cantor submitted to Crelle's Journal in 1877 was treated with suspicion by Kronecker, and only published after Dedekind intervened. Cantor resented Kronecker's opposition to his work and never again submitted papers to Crelle's journal.

Almost at the same time as the Cantor-Dedekind correspondence ended, Cantor began another important correspondence with Mittag-Leffler. Soon Cantor was publishing in Mittag-Leffler's journal *Acta Mathematica*. At the end of May, 1884, Cantor had the first recorded attack of depression. He recovered after a few weeks but now seemed less confident.

Mathematical worries began to trouble Cantor around 1884; in particular he began to worry that he could not prove the continuum hypothesis.

In 1897, Cantor discovered the first of the paradoxes in set theory. Cantor began a correspondence with Dedekind to try to understand how to solve the problems but recurring bouts of his mental illness forced him to stop writing. Whenever Cantor suffered from periods of depression he tended toward philosophy; he spent time trying to prove that Francis Bacon wrote the plays of Shakespeare.

A major event planned in Halle to mark Cantor's 70th birthday in 1915 had to be cancelled because of the war, but a smaller event was held in his home. In June, 1917 he entered a sanatorium for the last time and continually wrote to his wife asking to be allowed to go home. He died of a heart attack.

Notice that if $\text{card}(A) = \text{card}(B)$ via a function f_1 and $\text{card}(B) = \text{card}(C)$ via a function f_2, then $\text{card}(A) = \text{card}(C)$ via the function $f_2 \circ f_1$.

Definition 4.5.3 Let A and B be sets. If there is a one-to-one function from A to B but no bijection between A and B, then we will write

$$\text{card}(A) < \text{card}(B).$$

This notation is read "A has smaller cardinality than B."

We use the notation
$$\text{card}(A) \leq \text{card}(B)$$
to mean that either $\text{card}(A) < \text{card}(B)$ or $\text{card}(A) = \text{card}(B)$.

Notice that $\text{card}(A) \leq \text{card}(B)$ and $\text{card}(B) \leq \text{card}(C)$ imply that $\text{card}(A) \leq \text{card}(C)$. Moreover, if $A \subset B$, then the inclusion map $i(a) = a$ is a one-to-one function of A into B; therefore $\text{card}(A) \leq \text{card}(B)$.

EXAMPLE 4.5.4 Let $A = \{1, 2, 3\}$ and $B = \{2, 4, 6, 8, 10\}$. Then the function
$$f(x) = 2x \qquad \qquad .$$
is a one-to-one function from A to B. There is no one-to-one function from B to A (why not?). So we may write

$$\text{card}(A) < \text{card}(B).$$

Now let S be the integers $\mathbb{Z}$ and let T be the rational numbers $\mathbb{Q}$. Certainly the function
$$h(x) = x$$
is one-to-one from S to T. It is not clear (but see below) whether there is a one-to-one function from T to S. We may in any event write

$$\text{card}(S) \leq \text{card}(T).$$

$\diamond$

The next theorem gives a useful method for comparing the cardinality of two sets.

Theorem 4.5.5 (Cantor–Schroeder–Bernstein) *Let A, B, be sets. If there is a one-to-one function $f : A \to B$ and a one-to-one function $g : B \to A$, then A and B have the same cardinality.*

Remark 4.5.6 This remarkable theorem says that, if we can find an injection from A to B and an injection from B to A, then in fact there exists a single map that is a *bijection* of A to B. Observe that the two different injections may be completely unrelated. Often it is much easier to construct two separate injections than it is to construct a single bijection.

Proof: It is convenient to assume that A and B are disjoint; we may arrange this if necessary by replacing A by $\{(a,0) : a \in A\}$ and B by $\{(b,1) : b \in B\}$. Let D be the image of f and let C be the image of g. Let us define a *chain* to be a sequence of elements of either A or B—that is, a function $\phi : \mathbb{N} \to (A \cup B)$— such that

- $\phi(1) \in B \setminus D$;

- If for some j we have $\phi(j) \in B$, then $\phi(j + 1) = g(\phi(j))$;

- If for some j we have $\phi(j) \in A$, then $\phi(j + 1) = f(\phi(j))$.

We see that a chain is a sequence of elements of $A \cup B$ such that the first element is in $B \setminus D$, the second in A, the third in B, and so on. Obviously each element of $B \setminus D$ occurs as the first element of at least one chain.

Define $\mathcal{S} = \{a \in A : a$ is some term of some chain$\}$. It is helpful to note that

$$\mathcal{S} = \{x \in A : x \text{ can be written in the form}$$
$$x = g(f(g(\cdots f(g(y))\dots))) \text{ for some } y \in B \setminus D\}. \qquad (*)$$

Observe that $\mathcal{S} \subset C$.

We set

$$k(x) = \begin{cases} f(x) & \text{if } x \in A \setminus \mathcal{S} \\ g^{-1}(x) & \text{if } x \in \mathcal{S} \end{cases}$$

Note that the second half of this definition makes sense because $\mathcal{S} \subset C$ and because g is one-to-one. Then $k : A \to B$. We shall show that in fact k is a bijection.

First notice that f and g^{-1} are one-to-one. This is not quite enough to show that k is one-to-one, but we now reason as follows: If $f(x_1) = g^{-1}(x_2)$ for some $x_1 \in A \setminus \mathcal{S}$ and some $x_2 \in \mathcal{S}$, then $x_2 = g(f(x_1))$. But, by $(*)$, the fact that $x_2 \in \mathcal{S}$ now implies that $x_1 \in \mathcal{S}$. That is a contradiction. Hence k is one-to-one.

It remains to show that k is onto. Fix $b \in B$. We seek an $x \in A$ such that $k(x) = b$.

Case A: If $g(b) \in \mathcal{S}$, then $k(g(b)) \equiv g^{-1}(g(b)) = b$ hence the x that we seek is $g(b)$.

Case B: If $g(b) \notin \mathcal{S}$, then we claim that there is an $x \in A$ such that $f(x) = b$. Assume this claim for the moment.

Now the x that we just found must lie in $A \setminus \mathcal{S}$. For if not then x would be in some chain. Then $f(x)$ and $g(f(x)) = g(b)$ would also lie in that chain. Hence $g(b) \in \mathcal{S}$, and that is a contradiction. But $x \in A \setminus \mathcal{S}$ tells us that $k(x) = f(x) = b$. That completes the proof that k is onto. Hence k is a bijection.

To prove the claim that we made in Case B, notice that if there is no $x \in A$ with $f(x) = b$ then $b \in B \setminus D$. Thus some chain would begin at b. So $g(b)$ would be a term of that chain. Hence $g(b) \in S$ and that is a contradiction.

The proof of the Schroeder–Bernstein theorem is complete.   □

In what follows, we will consistently use some important and universally recognized terminology. An infinite set S is said to be *countable* if it has the same cardinality as the natural number $\mathbb{N}$. If an infinite set is *not* countable, that is if it does *not* have a bijection with the natural numbers $\mathbb{N}$, then it is said to be *uncountable*. Every infinite set is either countable or uncountable. One of our big jobs in this section of the book is to learn to recognize countable and uncountable sets.

Now it is time to look at some specific examples.

EXAMPLE 4.5.7 Let $\mathcal{E}$ be the set of all even integers and $\mathcal{O}$ the set of all odd integers. Then
$$\operatorname{card}(\mathcal{E}) = \operatorname{card}(\mathcal{O}).$$

Indeed, the function
$$f(j) = j + 1$$
is a bijection from $\mathcal{E}$ to $\mathcal{O}$.   ◇

EXAMPLE 4.5.8 Let $\mathcal{E}$ be the set of even integers. Then
$$\operatorname{card}(\mathcal{E}) = \operatorname{card}(\mathbb{Z}).$$

The function
$$g(j) = j/2$$
gives the bijection. Thus $\operatorname{card}(\mathcal{E}) = \operatorname{card}(\mathbb{Z})$.   ◇

This last example is a bit surprising, for it shows that a set (namely, $\mathbb{Z}$, the integers) can be put in one-to-one correspondence with a proper subset (namely $\mathcal{E}$, the even integers) of itself. In other words, $\mathbb{Z}$ has the same cardinality (that is, the same number of elements) as a proper subset of itself. This phenomenon is impossible for finite sets.

EXAMPLE 4.5.9 We have
$$\operatorname{card}(\mathbb{Z}) = \operatorname{card}(\mathbb{N}).$$

We define the function f from $\mathbb{Z}$ to $\mathbb{N}$ as follows:

- $f(j) = -(2j + 1)$ if j is negative

- $f(j) = 2j + 2$ if j is positive or zero

The values that f takes on the negative integers are $1, 3, 5, \ldots$, on the positive integers are $4, 6, 8, \ldots$, and $f(0) = 2$. Thus f is one-to-one and onto.   ◇

By putting together the preceding examples, we see that the set of even integers, the set of odd integers, and the set of all integers are countable sets.

EXAMPLE 4.5.10 The set of all ordered pairs of positive integers

$$S = \mathbb{N} \times \mathbb{N} = \{(j, k) : j, k \in \mathbb{N}\}$$

is countable.

To see this, we will use the Schroeder–Bernstein theorem. The function

$$f(j) = (j, 1)$$

is a one-to-one function from $\mathbb{N}$ to S. Also, the function

$$g(j, k) = j \cdot 10^{j+k} + k$$

is a function from S to $\mathbb{N}$. Let n be the number of digits in the number k. Notice that $g(j, k)$ is obtained by writing the digits of j, followed by $j + k - n$ zeros, then followed by the digits of k. For instance,

$$g(23, 714) = 23 \underbrace{000 \ldots 000}_{734} 714,$$

where there are $23 + 714 - 3 = 734$ zeros between the 3 and the 7. It is clear that g is one-to-one. By the Schroeder–Bernstein theorem, S and $\mathbb{N}$ have the same cardinality; hence S is countable. ◇

There are other ways to handle the last example, and we shall explore them in the exercises.

Since there is a bijection f of the set of *all* integers $\mathbb{Z}$ with the set $\mathbb{N}$, it follows from the last example that the set $\mathbb{Z} \times \mathbb{Z}$ of all pairs of integers (positive *and* negative) is countable. Let f be the function from Example 4.5.9. Then the map $(f \times f)(x, y) = (f(x), f(y))$ is a bijection of $\mathbb{Z} \times \mathbb{Z}$ to $\mathbb{N} \times \mathbb{N}$. Let h be the bijection, provided by Example 4.5.10, from $\mathbb{N} \times \mathbb{N}$ to $\mathbb{N}$. Then $h \circ (f \times f)$ is a bijection of $\mathbb{Z} \times \mathbb{Z}$ to $\mathbb{N}$.

Notice that the word "countable" is a good descriptive word: if S is a countable set, then we can think of S as having a first element s_1 (the one corresponding to $1 \in \mathbb{N}$), a second element s_2 (the one corresponding to $2 \in \mathbb{N}$), and so forth. Thus we write $S = \{s_1, s_2, \ldots\}$.

Definition 4.5.11 A set S is called *finite* if it is either empty or else there is a bijection of S with a set of the form $I_n \equiv \{1, 2, \ldots, n\}$ for some positive integer n. If S is not empty and if no such bijection exists, then the set is called *infinite*.

Remark 4.5.12 The empty set is a finite set, but not one of any particular interest. Nevertheless we must account for it, so we include it explicitly in the definition of "finite set."

In some treatments, a different approach is taken to the concepts of "finite" and "infinite" sets. In fact, one defines an infinite set to be one which can be put in one-to-one correspondence with a proper subset of itself. For instance, Example 4.5.8 shows that the set $\mathbb{Z}$ of all integers can be put in one-to-one correspondence with the set $\mathcal{E}$ of all even integers (and of course $\mathcal{E}$ is a proper subset of $\mathbb{Z}$). By contrast, a finite set *cannot* be put in one-to-one correspondence with a proper subset of itself. This last assertion amounts to verifying that I_k cannot be put in one-to-one correspondence with I_n when $k > n$. But any function $f : I_k \to I_n$ would be sending k letters to n mailboxes. By the pigeonhole principle, two letters would have to land in the same box. So the function cannot be one-to-one. Thus I_k and I_n do *not* have the same cardinality.

An important property of the natural numbers $\mathbb{N}$ is that any subset $S \subset \mathbb{N}$ has a least element. See the discussion in Section 4.2 and also later on in Section 5.1. This is known as the Well-Ordering Principle, and is studied in a course on logic (see also Section 5.1). In the present chapter, we take the properties of the natural numbers as given (see Section 5.1 for more on the natural numbers). We use some of these properties in the next proposition.

Proposition 4.5.13 *If S is a countable set and R is a subset of S, then either R is empty or R is finite or R is countable.*

Proof: Assume that R is not empty.

Write $S = \{s_1, s_2, \ldots\}$. Let j_1 be the least positive integer such that $s_{j_1} \in R$. Let j_2 be the least integer following j_1 such that $s_{j_2} \in R$. Continue in this fashion. If the process terminates at the nth step, then R is finite and has n elements.

If the process does not terminate, then we obtain an enumeration of the elements of R :

$$1 \longleftrightarrow s_{j_1}$$
$$2 \longleftrightarrow s_{j_2}$$
$$\cdots$$

etc.

All elements of R are enumerated in this fashion since $j_\ell \geq \ell$. Therefore R is countable. $\square$

A set is called *countable* if it is countably infinite, that is, if it can be put in one-to-one correspondence with the natural numbers $\mathbb{N}$. A set is called *denumerable* if it is either empty or finite or countable. In actual practice, mathematicians use the word "countable" to describe sets that are either finite or countable. In other words, they use the word "countable" interchangeably with the word "denumerable."

The set $\mathbb{Q}$ of all rational numbers consists of all expressions

$$\frac{a}{b},$$

where a and b are integers and $b \neq 0$. Thus $\mathbb{Q}$ can be identified with the set of all pairs (a, b) of integers. After discarding duplicates, such as $\frac{2}{4} = \frac{1}{2}$, and using the discussion following Examples 4.5.9 and 4.5.10 to the effect that $\mathbb{Z} \times \mathbb{Z}$ is countable, we find that the set $\mathbb{Q}$ is countable. We shall deal with the rational number system in a much more precise manner in Chapter 5.

Theorem 4.5.14 *Let S_1, S_2 be countable sets. Set $S = S_1 \cup S_2$. Then S is countable.*

Proof: Let us write

$$S_1 = \{s_1^1, s_2^1, \ldots\}$$
$$S_2 = \{s_1^2, s_2^2, \ldots\}.$$

If $S_1 \cap S_2 = \emptyset$, then the function

$$s_j^k \mapsto (j, k)$$

is a bijection of S with a subset of $\{(j, k) : j, k \in \mathbb{N}\}$. We proved earlier (Example 4.5.10) that the set of ordered pairs of elements of $\mathbb{N}$ is countable. By Proposition 4.5.13, S is countable as well.

If there exist elements which are common to S_1, S_2, then discard any duplicates. The same argument (use Proposition 4.5.10) shows that S is countable. $\square$

Proposition 4.5.15 *If S and T are each countable sets, then so is*

$$S \times T = \{(s, t) : s \in S, t \in T\}.$$

Proof: Since S is countable, there is a bijection f from S to $\mathbb{N}$. Likewise there is a bijection g from T to $\mathbb{N}$. Therefore the function

$$(f \times g)(s, t) = (f(s), g(t))$$

is a bijection of $S \times T$ with $\mathbb{N} \times \mathbb{N}$, the set of ordered pairs of positive integers. But we saw in Example 4.5.10 that the latter is a countable set. Hence so is $S \times T$. $\square$

Corollary 4.5.16 *If $S_1, S_2, \ldots, S_k$ are each countable sets, then so is the set*

$$S_1 \times S_2 \times \cdots \times S_k = \{(\mathbf{s_1}, \ldots, \mathbf{s_k}) : \mathbf{s_1} \in \mathbf{S_1}, \ldots, \mathbf{s_k} \in \mathbf{S_k}\}$$

consisting of all ordered k-tuples $(s_1, s_2, \ldots, s_k)$ with $s_j \in S_j$.

Proof: We may think of $S_1 \times S_2 \times S_3$ as $(S_1 \times S_2) \times S_3$. Since $S_1 \times S_2$ is countable (by the Proposition) and S_3 is countable, then so is $(S_1 \times S_2) \times S_3 = S_1 \times S_2 \times S_3$ countable. Continuing in this fashion (i.e., inductively), we can see that any finite product of countable sets is also a countable set. $\qquad\square$

Corollary 4.5.17 *The countable union of countable sets is countable.*

Proof: Let $A_1, A_2, \ldots$ each be countable sets. If the elements of A_j are enumerated as $\{a_k^j\}_{k=1}^\infty$ and if the sets A_j are pairwise disjoint, then the correspondence

$$a_k^j \longleftrightarrow (j, k)$$

is one-to-one between the union of the sets A_j and the countable set $\mathbb{N} \times \mathbb{N}$. This proves the result when the sets A_j have no common element. If some of the A_j have elements in common, then we discard duplicates in the union and use Proposition 4.5.13. $\qquad\square$

Proposition 4.5.18 *The collection $\mathcal{P}$ of all polynomials $p(x)$ with integer coefficients is countable.*

Proof: Let $\mathcal{P}_k$ be the set of polynomials of degree k with integer coefficients. A polynomial p of degree k having integer coefficients has the form

$$p(x) = p_0 + p_1 x + p_2 x^2 + \cdots + p_k x^k,$$

where the p_j are integer constants. The identification

$$p(x) \longleftrightarrow (p_0, p_1, \ldots, p_k)$$

identifies the elements of $\mathcal{P}_k$ with the $(k+1)$-tuples of integers. By Corollary 4.5.16, it follows that $\mathcal{P}_k$ is countable. But then Corollary 4.5.17 implies that

$$\mathcal{P} = \bigcup_{j=0}^{\infty} \mathcal{P}_j$$

is countable. $\qquad\square$

Definition 4.5.19 Let x be a real number. We say that x is *algebraic* if there is a polynomial p with integer coefficients such that $p(x) = 0$.

EXAMPLE 4.5.20 The number $\sqrt{2}$ is algebraic because it satisfies the polynomial equation $x^2 - 2 = 0$. The number $\sqrt{3} + \sqrt{2}$ is also algebraic. This assertion is less obvious, but in fact the number satisfies the polynomial equation $x^4 - x^2 + 1 = 0$. The numbers π and e are *not* algebraic, but this assertion is extremely difficult to prove. We say that π and e are *transcendental*. In the next proposition, we give an elegant method for showing that most real numbers are transcendental without actually saying what any of them are. ◇

Georg Cantor's remarkable discovery is that *not all infinite sets are countable*. We next give an example of this phenomenon.

In what follows, a *sequence* on a set S is a function from $\mathbb{N}$ to S. We usually write such a sequence as $s(1), s(2), s(3), \ldots$ or as $s_1, s_2, s_3, \ldots$.

EXAMPLE 4.5.21 There exists an infinite set which is not countable (we call such a set *uncountable*). Our example will be the set S of all sequences on the set $\{0, 1\}$. In other words, S is the set of all infinite sequences of 0's and 1's.

To see that S is uncountable, assume the contrary—that is, we assume that S is countable. Then there is a first sequence

$$\mathcal{S}^1 = \{s_j^1\}_{j=1}^\infty,$$

a second sequence

$$\mathcal{S}^2 = \{s_j^2\}_{j=1}^\infty,$$

and so forth. This will be a complete enumeration of all the members of S. But now consider the sequence $\mathcal{T} = \{t_j\}_{j=1}^\infty$, which we construct as follows:

- If $s_1^1 = 0$ then set $t_1 = 1$; if $s_1^1 = 1$ then set $t_1 = 0$;

- If $s_2^2 = 0$ then set $t_2 = 1$; if $s_2^2 = 1$ then set $t_2 = 0$;

- If $s_3^3 = 0$ then set $t_3 = 1$; if $s_3^3 = 1$ then set $t_3 = 0$;

$$\cdots$$

- If $s_j^j = 0$ then set $t_j = 1$; if $s_j^j = 1$ then set $t_j = 0$;

etc.

Now the sequence $\mathcal{T} = \{t_j\}$ differs from the first sequence $\mathcal{S}^1$ in the first element: $t_1 \neq s_1^1$.

The sequence $\mathcal{T}$ differs from the second sequence $\mathcal{S}^2$ in the second element: $t_2 \neq s_2^2$.

And so on: the sequence $\mathcal{T}$ differs from the jth sequence $\mathcal{S}^j$ in the jth element: $t_j \neq s_j^j$. So the sequence $\mathcal{T}$ is not in the set S. But $\mathcal{T}$ is *supposed* to be in the set S because it is a sequence of 0's and 1's and all of these are supposed to have been enumerated in our enumeration of S.

This contradicts our assumption, so S must be uncountable. ◊

EXAMPLE 4.5.22 Consider the set of all decimal representations of numbers strictly between 0 and 1—both terminating and nonterminating. Here a terminating decimal is one of the form

0.43926

while a nonterminating decimal is one of the form

$$0.14159265\ldots.$$

In the case of the nonterminating decimal, no repetition is implied; the decimal simply continues without cease.

Now the set of all those decimals containing only the digits 0 and 1 can be identified in a natural way with the set of sequences containing only 0 and 1 (just put commas between the digits). And we just saw that the set of such sequences is uncountable.

Since the set of all decimal numbers is an even bigger set, it must be uncountable also. [Put a different way, if the set of all decimal numbers *were* countable, then any of its infinite subsets would be countable—that is the content of Proposition 4.5.13. Thus the collection of decimal numbers containing only the digits 0 and 1 would be countable, and that is a contradiction.]

As you may know, the set of all decimals identifies with the set of all real numbers. [Many real numbers have two decimal representations—one terminating and one not. Think for a moment about which numbers these are, and why this observation does not invalidate the present discussion.] We find then that the set $\mathbb{R}$ of all real numbers is uncountable. (Contrast this with the situation for the rationals.) In Chapter 6 we will learn more about how the real number system is constructed using just elementary set theory.   ◇

Proposition 4.5.23 *The set of all algebraic real numbers is countable. The set of all transcendental numbers is uncountable.*

Proof: Let $\mathcal{P}$ be the collection of all polynomials with integer coefficients. We have already noted in Proposition 4.5.18 that $\mathcal{P}$ is a countable set. If $p \in \mathcal{P}$ then let s_p denote the set of real roots of p. Of course s_p is finite, and the number of elements in s_p does not exceed the degree of p. Then the set A of algebraic real numbers may be written as

$$A = \cup_{p \in \mathcal{P}}\, s_p\,.$$

This is the countable union of finite sets, so of course it is countable.

Now that we know that the set of algebraic numbers is countable, we can notice that the set T of transcendental numbers must be uncountable. For $\mathbb{R} = A \cup T$. If T were countable then, since A is countable, it would follow that $\mathbb{R}$ is countable. But that is not so.   □

Our last result in this section is a counterpoint to Proposition 4.5.13 and the discussion leading up to it.

Proposition 4.5.24 *Let S be any infinite set. Then S has a subset T that is countable.*

Proof: Let $t_1 \in S$ be any element. Now let $t_2 \in S$ be any element that is distinct from t_1. Continue this procedure. It will not terminate, because that would imply that S is finite. And it will produce a countable set T that is a subset of S. □

To repeat the main point of this section, the natural numbers have a cardinality that we call *countable*, and the real numbers have a cardinality that we call *uncountable*. These cardinalities are distinct. In fact the real numbers form a larger set because there is an injective mapping of the natural numbers into the reals but not the other way around. We refer to the cardinality of the natural numbers as "countable" and to that of the real numbers as "the cardinality of the continuum."

It is natural to ask whether there is a set with cardinality strictly between countable and the continuum. Georg Cantor posed this question one hundred years ago, and his failed attempts to resolve the question tormented his final years. We shall discuss the resolution of this "continuum hypothesis" in the Appendix, Subsection A2.2. See also

https://mathworld.wolfram.com/ContinuumHypothesis.html .

It is an important result of set theory (due to Cantor) that, given any set S, the set of all subsets of S (called the *power set* of S) has strictly greater cardinality than the set S itself. As a simple example, let $S = \{a, b, c\}$. Then the set of all subsets of S is

$$\{\emptyset, \{a\}, \{b\}, \{c\}, \{a, b\}, \{a, c\}, \{b, c\}, \{a, b, c\}\}.$$

The set of all subsets has eight elements while the original set has three.

We stress that this result is true not just for finite sets but also for infinite sets: if S is an infinite set then the set of all its subsets (the power set) has greater cardinality than S itself. Thus there are infinite sets of arbitrarily large cardinality. In other words, there is no "greatest" cardinal. This fact is so important that we now formulate it as a theorem.

Theorem 4.5.25 (Cantor) *Let S be any set. Then the power set $\mathcal{P}(S)$, consisting of all subsets of S, has cardinality greater than the cardinality of S. In other words,*

$$\mathrm{card}(S) < \mathrm{card}(\mathcal{P}(S)).$$

Proof: First observe that the function

$$
\begin{aligned}
f : S &\longrightarrow \mathcal{P}(X)\\
s &\longmapsto \{s\}
\end{aligned}
$$

is one-to-one. Thus we see that $\mathrm{card}(S) \leq \mathrm{card}(\mathcal{P}(S))$. We need to show that there is no function from S *onto* $\mathcal{P}(S)$. Let $g : S \to \mathcal{P}(S)$. We will produce an element of $\mathcal{P}(S)$ that cannot be in the image of this mapping.

Define $T = \{s \in S : s \notin g(s)\}$. Assume, seeking a contradiction, that $T = g(z)$ for some $z \in S$. By definition of T, the element $z \in T$ if and only if $z \notin g(z)$; thus $z \in T$ if and only if $z \notin T$. That is a contradiction. We see that g cannot map S onto $\mathcal{P}(S)$, therefore $\text{card}(S) < \text{card}(\mathcal{P}(S))$. $\square$

In some of the examples in this section, we constructed a bijection between a given set (such as $\mathbb{Z}$) and a proper subset of that set (such as $\mathcal{E}$, the even integers). It follows from the definitions that this is possible only when the sets involved are infinite. In fact any infinite set can be placed in a set-theoretic isomorphism with a proper subset of itself. We explore this assertion in the exercises.

Put in other words, we have come upon an intrinsic characterization of infinite sets. We state it (without proof) as a proposition:

Proposition 4.5.26 *Let S be a set. The set S is infinite if and only if it can be put in one-to-one correspondence with a proper subset of itself.*

Exercise 4.41 outlines a proof of this proposition.

Paul Joseph Cohen
1934–2007

Paul was the youngest of his parents' four children and he was brought up in Brooklyn, New York. He began to study advanced mathematics from a young age. He attended Stuyvesant High School in New York City, graduating in 1950 at the young age of sixteen years. Continuing to study at Chicago for his doctorate under the supervision of Antoni Zygmund, he was awarded his Ph.D. in 1958 for his doctoral thesis "Topics in the Theory of Uniqueness of Trigonometric Series." John Thompson was one such fellow research student at Chicago. Cohen, through these friendships, had also begun to take an interest in logic.

In 1957, before the award of his doctorate, Cohen was appointed as an Instructor in Mathematics at the University of Rochester for a year. He then spent the academic year 1958–1959 at the Massachusetts Institute of Technology before spending 1959–1961 as a fellow at the Institute for Advanced Study at Princeton. In 1961, Cohen was appointed to the faculty at Stanford University as an Assistant professor of mathematics. He was promoted to Associate professor in mathematics in the following year and, also in 1962, was awarded an Alfred P. Sloan research fellowship. In August 1962, Cohen participated in the International Congress of Mathematicians in Stockholm.

He was promoted to full Professor at Stanford University in 1964 having, by this time, solved one of the most challenging open problems in mathematics. Cohen used a technique called "forcing" to prove the independence in set theory of the Axiom of Choice and of the generalized continuum hypothesis.

Kurt Gödel wrote to him: Let me repeat that it is really a delight to read your proof of the independence of the continuum hypothesis. I think that in all essential respects you have given the best possible proof and this does not happen frequently. Reading your proof had a similarly pleasant effect on me as seeing a really good play.

In 1966 Cohen published the monograph *Set Theory and the Continuum Hypothesis* based on a course he gave at Harvard in spring 1965.

In addition to his work on set theory, Cohen worked on differential equation and harmonic analysis. Peter Sarnak (Professor of mathematics at Princeton and a former doctoral student of Cohen's with the thesis "Prime Geodesic Theorems" (1980)) says of Cohen that "Paul Cohen was one of the most brilliant mathematicians of the 20th century. Like many great mathematicians, his mathematical interests and contributions were very broad, ranging from mathematical analysis and differential equations to mathematical logic and number theory. This breadth was highlighted in a conference held at Stanford last September celebrating Cohen's work and his 72nd birthday. The gathering consisted of leading experts in different fields who normally would not find themselves listening to the same set of lectures ... " Cohen was a dynamic and enthusiastic lecturer and teacher. He made mathematics look simple and unified. He was always eager to share his many ideas and insights in diverse fields. His passion for mathematics never waned.

4.6   Hilbert's Hotel Infinity

This somewhat frivolous, but absolutely mathematically rigorous, section gives a nice illustration of the concept of countable set.

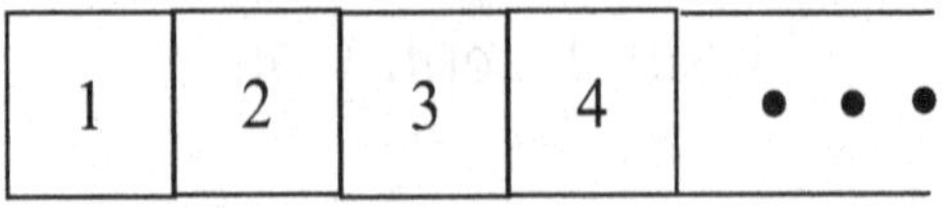

FIGURE 4.1
The hotel infinity.

In 1934, the eminent mathematician David Hilbert, in a public lecture, proposed the idea of "Hotel Infinity." It is a fascinating notion, and we describe it here. The charming and entertaining book [GAM] explores the idea further.

Imagine a hotel with infinitely many rooms. For convenience, these rooms are numbered $1, 2, 3, \ldots$. See Figure 4.1.

Now suppose that, on a given evening, all the rooms are occupied. But a new guest, who does *not* have a reservation, shows up. What is the hotel manager to do?

If this were an ordinary hotel with finitely many rooms, then the manager would be stuck. She would have no room for the new guest, and would have to send her away. But instead she has infinitely many rooms, so she can do something creative:

> She moves the guest in Room 1 to Room 2, and the guest in Room 2 to Room 3, and the guest in Room 3 to Room 4, and so on. In this way every one of the pre-existing guests will still have a room. But Room 1 will be freed up and she can put the new guest in Room 1.

Miraculous, no??

The same thing would work if 5 new guests were to show up. Here is how she would proceed:

> She moves the guest in Room 1 to Room 6, and the guest in Room 2 to Room 7, and the guest in Room 3 to Room 8, and so on. In this way every one of the pre-existing guests will still have a room. But Rooms 1 through 5 will be freed up and she can put the new guests in Rooms 1 through 5.

Now what if infinitely many new guests show up? That is more of a challenge. But the hotel manager is a talented mathematical thinker, and she comes up with this solution:

> She moves the guest in Room 1 to Room 2. Then she moves the guest in Room 2 to Room 4. Next she moves the guest in Room 3 to Room 6. In general, she moves the guest in Room k to Room $2k$. This frees up Rooms $1, 3, 5, 7, \ldots$. So infinitely many rooms have been freed up, and she can accommodate infinitely many new guests.

The devil is working to try to trip up our intrepid hotel manager. So he comes up with the following situation. He brings in infinitely many buses, and each bus has infinitely many passengers. He wants her to provide accommodations for *all* of these new guests. How does she do it?

First, she moves the guest in Room 1 to Room 2. Then she moves the guest in Room 2 to Room 4. Next she moves the guest in Room 3 to Room 6. And so on.

This frees up all the odd-numbered rooms. But she needs to accommodate infinitely many collections of infinitely many people. How can she do it?

She takes the people from the first bus and puts them in the rooms numbered
$$3^1, 3^2, 3^3, \ldots$$
So the first bus passengers all go into rooms whose numbers are powers of 3. These are of course odd-numbered rooms, so they are vacant.

Next, she takes the people from the second bus and puts them in the rooms numbered
$$5^1, 5^2, 5^3, \ldots$$
So the second bus passengers all go into rooms whose numbers are powers of 5. These are of course odd-numbered rooms, and they are different from the rooms with numbers that are powers of 3. So these rooms are vacant, and can accommodate the passengers from bus number 2.

For the next step, she takes the people from the third bus and puts them in the rooms numbered
$$7^1, 7^2, 7^3, \ldots$$
So the third bus passengers all go into rooms whose numbers are powers of 7. These are of course odd-numbered rooms, and they are different from the rooms with numbers that are powers of 3 or of 5. So these rooms are vacant, and can accommodate the passengers from bus number 3.

The hotel manager continues in this fashion, placing passengers from bus k into rooms with numbers that are powers of the kth odd prime number.

Take particular note that, for the fourth bus, we do not put the guests into the rooms numbered 9^1, 9^2, 9^3, ... Because these are actually powers of 3, so these rooms would have been filled when we accommodate the guests from the first bus. It is important that we only use rooms with numbers that are *powers of the prime numbers*. So, for the fourth bus, we use the rooms with numbers 11^1, 11^2, 11^3, ...

The method that we have described for infinitely many buses each with infinitely many passengers certainly works, but it also leaves infinitely many rooms empty. For example, the room with number $15 = 5 \cdot 3$ will be empty. But this may be convenient for the hotel.

In fact Georg Cantor (1845–1918) was the man who developed techniques for handling infinite sets. He showed that infinite sets come in different sizes, and he created a calculus for manipulating infinite sets. He is remembered today as one of the great mathematicians of the last hundred years. Hilbert's

Hotel Infinity was in fact a commentary on and an appreciation of Cantor's work.

Exercise: Imagine infinitely many buses showing up, each with infinitely many passengers, but these people are only willing to stay in rooms with even numbers. How can the hotel manager accommodate these new guests?
See

https://owlcation.com/stem/Hilberts-Paradox-of-the-Grand-Hotel-Another-Look-at-Infinity

for more about the hotel infinity and

https://mathshistory.st-andrews.ac.uk/Biographies/Hilbert

for more about Hilbert.

Exercises

1. Consider the relation on $\mathbb{Z}$ defined by $(m, n) \in \mathcal{R}$ if $m + n$ is even. Prove that this is an equivalence relation. What are the equivalence classes?

2. Consider the relation on $\mathbb{Z} \times (\mathbb{Z} \setminus \{0\})$ defined by $(m, n)\mathcal{R}(m', n')$ provided that $m \cdot n' = m' \cdot n$. Prove that this is an equivalence relation. Can you describe the equivalence classes?

3. Consider the relation on $\mathbb{Z} \times \mathbb{Z}$ defined by $(m, n)\mathcal{R}(m', n')$ provided that $m + n' = m' + n$. Prove that this is an equivalence relation. Can you describe the equivalence classes? Can you pick a representative for each equivalence class that will help to exhibit what the equivalence relation is?

4. Consider the relation defined on the cartesian plane by $(x, y)\mathcal{R}(x', y')$ if $y = y'$. Prove that this is an equivalence relation. Can you describe the equivalence classes? Can you pick a representative for each equivalence class that will help to exhibit what the equivalence relation is?

5. Consider the relation defined on the cartesian plane by $(x, y)\mathcal{R}(x', y')$ if $y - y'$ is an integer and $x - x'$ is an integer. Prove that this is an equivalence relation. Can you describe the equivalence classes? Can you pick a representative for each equivalence class that will help to exhibit what the equivalence relation is?

6. Consider the relation defined on the collection of all circles in the Euclidean plane by $C_1 \mathcal{R} C_2$ if the circle C_1 and the circle C_2 have the same center. Prove that this is an equivalence relation. Can you describe the equivalence classes? Can you pick a representative for each equivalence class that will help to exhibit what the equivalence relation is?

7. Let S be the set of all living people. Let $x, y \in S$. Say that x is related to y if x and y have some blood relation in common. Is this an equivalence relation? Why or why not?

8. Consider the relation on $\mathbb{Q} \times (\mathbb{Q} \setminus \{0\})$ defined by $(m, n)\mathcal{R}(m', n')$ provided $m \cdot n' = m' \cdot n$. Prove that this is an equivalence relation. Can you describe the equivalence classes? Why is the outcome in this exercise different from that in Exercise 4.2?

9. Let $S = \{a, b, c, d\}$ and $T = \{1, 2, 3, 4, 5, 6, 7\}$. Which of the following relations on $S \times T$ is a function?

   (a) $\{(a, 4), (d, 3), (c, 3), (b, 2)\}$
   (b) $\{(a, 5), (c, 4), (d, 3)\}$
   (c) $\{(a, 1), (b, 1), (c, 1), (d, 1)\}$
   (d) $\{(a, 2), (b, 2), (c, 3), (d, 3)\}$
   (e) $\{(d, 1), (c, 2), (b, 3), (a, 4)\}$
   (f) $\{(d, 7), (c, 6), (c, 5), (a, 4), (b, 2)\}$
   (g) $\{(a, 6), (c, 9)\}$

10. Which of the following functions is one-to-one? Which is onto?

   (a) $f : \mathbb{N} \to \mathbb{N}$ $f(m) = m + 2$
   (b) $g : \mathbb{Z} \to \mathbb{Z}$ $g(m) = 2m - 7$
   (c) $h : \mathbb{R} \to \mathbb{R}$ $h(x) = x - x^3$
   (d) $f : \mathbb{Q} \to \mathbb{Q}$ $f(x) = x^2 + 4x$
   (e) $g : \mathbb{N} \to \mathbb{N}$ $g(n) = n(n + 1)$
   (f) $h : \mathbb{R} \to \mathbb{R}$ $h(n) = +\sqrt{n^2 + 1}$
   (g) $f : \mathbb{Z} \to \mathbb{N}$ $f(n) = n^2 + n + 1$
   (h) $g : \mathbb{N} \to \mathbb{Z}$ $g(k) = k^3 + 2k$
   (i) $h : \mathbb{N} \to \mathbb{Q}$ $h(t) = t/(t + 1)$
   (j) $f : \mathbb{Q} \to \mathbb{Q}$ $f(y) = y^2 - y$

11. Consider all ordered triples of positive integers. If $\alpha = (a, b, c)$ and $\alpha' = (a', b', c')$ are two such triples, then we say that $\alpha < \alpha'$ if either

   (a) $a < a'$
   (b) $a = a'$ and $b < b'$

   or

(c) $a = a'$, $b = b'$, and $c < c'$.

Discuss, in the language of Section 4.2, what type of order relation this is. This ordering is called the *lexicographic ordering*. In view of the way that we order words in a dictionary, explain why the ordering just described deserves that name.

12. Explain how the ordering described in Exercise 5.11 can be generalized to ordered k-tuples $(a_1, \ldots, a_k)$ of positive integers.

13. Consider the set S of all infinite sequences $\{a_1, a_2, \ldots\}$ of real numbers. Say that two such sequences $\alpha = \{a_1, a_2, \ldots\}$ and $\alpha' = \{a'_1, a'_2, \ldots\}$ satisfy $\alpha \leq \alpha'$ if the terms of α are eventually less than or equal to the terms of α'. This means that there exists a $K > 0$ such that $a_j \leq a'_j$ for all $j \geq K$. Discuss, in the language of Section 4.2, what sort of order relation this is.

14. Consider a relation defined on all living people defined by $a \mathcal{R} b$ if a and b are of the same sex and a is strictly younger than b. Is this an equivalence relation? Does it fit one of the orderings described in Section 4.2?

15. Express parts (b), (c), (d) of Definition 4.4.1 using the language of ordered pairs. Imitate our discussion of part (a) in the text.

* 16. Classify the Pythagorean triples. That is, find all triples m, n, p of positive integers such that $m^2 + n^2 = p^2$. [**Hint:** Consider m, n, p of the form $m = s^2 - t^2$, $n = 2st$, $p = s^2 + t^2$, for s, t positive integers with no common divisors.]

17. Give an example of a function $f : \mathbb{N} \to \mathbb{Z}$ that is onto.

18. Give an example of a function $f : \mathbb{Q} \to \mathbb{N}$ that is onto.

19. Prove that there is no function $g : \mathbb{N} \to \mathbb{R}$ that is onto.

20. What is the cardinality of each of the following sets?

    (a) $\mathbb{N} \times \mathbb{Q}$
    (b) $\mathbb{N} \times \mathbb{N}$
    (c) $\mathbb{R} \times \mathbb{Q}$
    (d) $\mathcal{P}(\mathbb{Q})$
    (e) $\mathbb{C}$
    (f) $\mathbb{R} \setminus \mathbb{N}$
    (g) $\mathbb{Q} \setminus \mathbb{N}$
    (h) The set of all decimal expansions, terminating or nonterminating, that include only the digits 3 and 7.
    (i) The set of all *terminating* decimal expansions that include only the digits 3 and 7.
    (j) The set of all solutions of all quadratic polynomials with integer coefficients.

(k) The set of all solutions of all quadratic polynomials with real coefficients.

(l) The set of all subsets of $\mathbb{N}$ that have at least three and not more than eight elements.

(m) The set of all subsets of $\mathbb{Z}$ with at least six elements.

* **21.** Let S and T be sets and let $f : S \to T$ and $g : T \to S$ be arbitrary functions. Prove that there is a subset $A \subset S$ and a subset $B \subset T$ such that $f(A) = B$ and $g(T \setminus B) = S \setminus A$.

* **22.** Find all functions $f : \mathbb{R} \setminus \{0\} \to \mathbb{R} \setminus \{0\}$ that are one-to-one and onto and such that $f^{-1}(x) = 1/f(x)$.

* **23.** Let $f : \mathbb{R} \to \mathbb{R}$ be a function with the property that every x in $\mathbb{R}$ is a local minimum. That is, for $x \in \mathbb{R}$ there is an $\epsilon_x > 0$ so that if $t \in (x - \epsilon_x, x + \epsilon_x)$, then $f(t) \geq f(x)$. Then prove that the image of f is countable.

24. Construct an onto function $f : \mathbb{R} \to \mathbb{R} \times \mathbb{R}$. Use this function to equip $\mathbb{R} \times \mathbb{R}$ with an ordering.

25. Find the domain and image of each of these relations:

(a) $\{(x, y) \in \mathbb{R} \times \mathbb{R} : x = \sqrt{y + 3}\}$

(b) $\{(x, y) \in \mathbb{Q} \times \mathbb{Q} : y = 1/(x^2 - 4)\}$

(c) $\{(\alpha, \beta) : \alpha$ is a person, β is a person, and α is the father of $\beta\}$

(d) $\{(\alpha, \beta) : \alpha$ is a person, β is a person, and α is a parent of $\beta\}$

(e) $\{(x, y) \in \mathbb{Q} \times \mathbb{Q} : x^2 + y^2 < 1\}$

(f) $\{(x, y) \in \mathbb{N} \times \mathbb{Q} : x \cdot y$ is an integer$\}$

(g) $\{(x, y) \in \mathbb{R} \times \mathbb{R} : x \cdot y$ is rational$\}$

(h) $\{(x, y) \in \mathbb{Z} \times \mathbb{Z} : x - y = 2\}$

26. We have used the phrase "ordered pair" in this book without giving a precise definition of the phrase. As noted in the text, we *could* define the ordered pair (a, b) to be the set $\{\{a\}, \{a, b\}\}$. This is clearly distinct from the ordered pair (b, a), which would be $\{\{b\}, \{b, a\}\} = \{\{b\}, \{a, b\}\}$. Using this technical definition, *prove* that $(a, b) = (a', b')$ if and only if $a = a'$ and $b = b'$.

27. Give a rigorous definition of "ordered triple" based on the ideas in Exercise 4.26.

28. Declare two real numbers to be related if their difference is rational. Prove that this is an equivalence relation. How many elements are in each equivalence class? How many equivalence classes are there?

29. Formulate a notion of composition of two relations. Formulate a notion of the inverse of a relation. Now express the ideas of reflexivity, symmetry, and transitivity for a relation in the language of inverse and composition of relations.

30. Let $A = \{1, 2, 3\}$, $B = \{a, b, c\}$, $C = \{s, t, u\}$. Define functions

$$\begin{aligned} f &= \{(1, c), (2, c), (3, a)\} \\ g &= \{(a, t), (b, s), (c, u)\} \end{aligned}$$

What are the domain and image of f? What are the domain and image of g? Calculate $g \circ f$ and g^{-1}.

31. Give precise meaning to, and prove, the statement that the intersection of two functions is a function. Is it also the case that the union of two functions is a function?

32. Give an explicit example of a function $f : \mathbb{Q} \to \mathbb{Q}$ such that f is one-to-one and onto but such that $f(x) > x^3$ for every x.

33. Let $f : \mathbb{R} \to \mathbb{R}$ be a function. Suppose that, for every surjective function $g : \mathbb{R} \to \mathbb{Z}$, it holds that $g \circ f$ is surjective. Then prove that f is surjective.

34. Let X be a set such that there exists a surjective function $f : X \to \mathbb{Z}$. Then prove that X is infinite.

35. Let X be a set such that there exists a surjective function $f : X \to \mathbb{R}$. Then prove that X is uncountable.

36. Consider the following relation on $\mathbb{N}$: $(x, y) \in \mathcal{R}$ if $x < y + 2$. What sort of order relation is $\mathcal{R}$?

37. Consider the following relation on $\mathbb{N}$: $(x, y) \in \mathcal{R}$ if $x < y - 2$. What sort of order relation is $\mathcal{R}$?

38. Let S be the set of all living people. Tell which of the following are equivalence relations on S. Give detailed reasons for your answers.

    (a) x is related to y if x and y are siblings
    (b) x is related to y if y is presently a spouse of x
    (c) x is related to y if y has at one time or another been a spouse of x
    (d) x is related to y if y is a parent of x
    (e) x is related to y if y is a child of x
    (f) x is related to y if x hates y but y loves x
    (g) x is related to y if x hates y and y hates x
    (h) x is related to y if x and y have a common ancestor

39. Let S be the collection of all polynomials with real coefficients. Say that $p, q \in S$ are related if the number 0 is a root of $p - q$. Is this an equivalence relation on S?

40. Redo Example 4.5.10 by laying out the ordered pairs in a tableau (see below) and counting them along diagonals that proceed from the lower left to the upper right.

$$(1,1) \quad (1,2) \quad (1,3) \quad (1,4) \quad \cdots$$
$$(2,1) \quad (2,2) \quad (2,3) \quad (2,4) \quad \cdots$$
$$(3,1) \quad (3,2) \quad (3,3) \quad (3,4) \quad \cdots$$
$$\cdots$$

* 41. In this exercise we indicate why any infinite set can be put into a set-theoretic isomorphism with a proper subset of itself. You should provide the details.

Let S be a set with infinitely many elements. By Proposition 4.5.24, there is a subset $T \subset S$ that is countable. Now we know that the set of natural numbers can be placed in one-to-one correspondence with a proper subset of itself, hence so can T be placed in one-to-one correspondence with a proper subset T' of itself. Let g denote the set-theoretic isomorphism of T with T'.

Now set $U = S \setminus T$. We may construct a set-theoretic isomorphism of $S = U \cup T$ to $U \cup T'$ (a proper subset of S) by following these rules:

(a) Map each element of U to itself;

(b) Map each element $t \in T$ to $g(t) \in T'$.

42. TRUE or FALSE: If $S_1 \supset S_2 \supset \cdots$ are each uncountable sets then $\cap_j S_j$ is a non-empty, indeed an uncountable, set.

43. Let A and B be sets. Let us say that A and B are related if there exists a set-theoretic isomorphism from A to B. Prove that this is an equivalence relation. Each equivalence class is called a *cardinal number*.

44. Say that two real numbers x and y are related if there is an integer k such that $k < x \leq k+1$ and $k < y \leq k+1$. Explain why this is an equivalence relation. Draw a figure that shows the equivalence classes in the real line.

45. Say that two real numbers are related if the first five digits of each of their decimal expansions (the five digits to the right of the decimal point) are equal. After giving a precise formulation of this relation, show that it is an equivalence relation. Give a verbal description of each equivalence class.

46. Give an explicit example of a function $f : \mathbb{Q} \to \mathbb{Q}$ such that f is one-to-one and onto but such that $f(x) \neq x$ for every x.

47. Write $\mathbb{N} \times \mathbb{N}$ as the countable union of finite sets. Conclude that $\mathbb{N} \times \mathbb{N}$ is countable.

48. Write $\mathbb{N} \times \mathbb{N}$ as the finite union of countable sets. Conclude that $\mathbb{N} \times \mathbb{N}$ is countable.

49. Let S and T be sets. We let T^S denote the set of all functions from S to T. For the specific example $S = \{1, 2, 3\}$ and $T = \{a, b\}$, write out the set T^S. Also write out the set S^T.

50. Refer to Exercise 4.49 for notation. Suppose that the set S has k elements and the set T has m elements, with k, m positive integers. What does the set S^T have to do with the number k^m?

51. Refer to Exercise 4.49 for notation. Let S be a finite set and let T be a set with three elements. Prove that there is a one-to-one correspondence between S^T and $S \times S \times S$.

52. Refer to Exercise 4.49 for notation. Let $S = \{1, 2\}$, $T = \{a, b, c\}$, and $U = \{\alpha, \beta\}$. Prove that there is a natural one-to-one correspondence between $(S^T)^U$ and $S^{T \times U}$.

53. Refer to Exercise 4.49 for notation. Let S be a set and let T be the empty set $\emptyset$. What can you say about S^T and T^S?

54. Exhibit a one-to-one correspondence of the set $\mathbb{R}$ of real numbers with a proper subset of $\mathbb{R}$. [**Hint:** Refer to Exercise 4.41.]

55. Write the set of all sequences of 0's and 1's using the notation introduced in Exercise 4.49.

56. Refer to Example 4.1.3. Let $S = \mathbb{N}$ and $T = \mathbb{N}$. Define a relation $\mathcal{R}$ on S and T by the condition $(s, t) \in \mathcal{R}$ if $s^2 + t^2$ is itself a perfect square. Show that the domain of $\mathcal{R}$ has infinitely many elements (Exercise 4.16 above may be of some help). Show that the image of $\mathcal{R}$ has infinitely many elements.

57. Which of these relations is a partial order? Give a concrete reason for your answer in each case.

(a) $\mathcal{P}$ is the collection of all polynomials with integer coefficients. We say that $p\mathcal{R}q$ if the degree of p is less than or equal to the degree of q.

(b) $\mathbb{Q}$ is the set of all positive rational numbers in lowest terms (i.e., the numerator and denominator have no factors in common). We say that $q_1\mathcal{R}q_2$ if the denominator of q_1 is less than the denominator of q_2.

(c) $\mathbb{R}$ is the set of real numbers. We say that $r_1\mathcal{R}r_2$ if the integer part of r_1 is less than the integer part of r_2.

(d) $\mathcal{S}$ is the collection of all subsets of the real numbers. We say that $A\mathcal{R}B$ if A has fewer elements than B.

5

Number Systems

In this chapter we treat all the basic number systems. This includes the natural numbers, the integers, the rational numbers, the real numbers, the complex numbers, and beyond. We do this rigorously—certainly more rigorously than anything you have ever seen before.

You will see right away in Section 5.1 that we begin with an axiom system for the natural numbers. You will have to read carefully to get the most from this treatment.

5.1 The Natural Number System

Giuseppe Peano's axioms for the natural numbers are as follows. In this discussion, we will follow tradition and use the notation $'$ to denote the "successor" of a natural number. For instance, the successor of 2 is $2'$. Intuitively, the successor of n is the number $n + 1$. However addition is something that comes later; so we formulate the basic properties of the natural numbers in terms of the successor function.

PEANO'S AXIOMS FOR THE NATURAL NUMBERS

P1 $1 \in \mathbb{N}$.

P2 If $n \in \mathbb{N}$, then $n' \in \mathbb{N}$.

P3 There is no natural number n such that $n' = 1$.

P4 If m and n are natural numbers and if $m' = n'$, then $m = n$.

P5 Let P be a property. If

  (i) $P(1)$ is true;

  (ii) $P(j) \Rightarrow P(j')$ for every $j \in \mathbb{N}$

then $P(n)$ is true for every $n \in \mathbb{N}$.

As Suppes says in [SUP, pp. 121 ff.], these axioms for the natural numbers are almost universally accepted (although E. Nelson [NEL], among others, has

DOI: 10.1201/9781003214564-5

found it useful to explore how to develop arithmetic without Axiom **P5**). They have evolved into their present form so that the natural numbers will satisfy those properties that are generally recognized as desirable. Let us briefly mention what each of the axioms signifies:

P1 asserts that $\mathbb{N}$ contains a distinguished element that we denote by 1.

P2 asserts that each element of $\mathbb{N}$ has a successor.

P3 asserts that 1 is not the successor of any natural number; in other words, 1 is in a sense the "first" element of $\mathbb{N}$.

P4 asserts that if two natural numbers have the same successor then they are in fact the same natural number.

P5 asserts that the method of mathematical induction is valid.

Some obvious, and heuristically appealing, properties of the natural numbers can be derived rather directly from Peano's axioms. Here is an example:

Proposition 5.1.1 *Let n be a natural number other than 1. Then $n = m'$ for some natural number m.*

Remark 5.1.2 This proposition makes the intuitively evident assertion that every natural number, except 1, is a successor. Another way of saying this is that every natural number except 1 has a predecessor. This claim is clearly not an explicit part of any of the five axioms. In fact the only axiom that has a hope of implying the proposition is the inductive axiom, as we shall now see.

Proof: Let $P(j)$ be the statement "either $j = 1$ or $j = n'$ for some natural number n."

Clearly $P(j)$ is true when $j = 1$.

Now suppose that the statement $P(j)$ has been established for some natural number j. We wish to establish it for j'. But j' is, by definition, a successor. So the statement is true.

This completes our mathematical induction.   □

This proof is misleading in its simplicity. The proof consists of little more than *interpreting* axiom **P5**. Some other desirable properties of the natural numbers are much more difficult to prove directly from the axioms. As an instance, to prove that there is no natural number lying between k and k' is complicated. Indeed, the entire concept of ordering is extremely tricky. And the single most important property of the natural numbers (one that is essentially equivalent to the mathematical induction axiom, as we shall later see) is that the natural numbers are well ordered in a canonical fashion. So we must find an efficient method for establishing the properties of the natural numbers that are connected with order.

It is generally agreed (see [SUP, p. 121 ff.]) that the best way to develop further properties of the natural numbers is to treat a specific model. Even that approach is nontrivial; so we shall only briefly sketch the construction of a model and further sketch its order properties so that we may discuss well ordering. The approach that we take is by way of the so-called *finite ordinal arithmetic*.

Let us define a model for the natural numbers as follows:

$$1 = \{\emptyset\}$$
$$2 = 1 \cup \{1\} = \{\emptyset, \{\emptyset\}\}$$
$$3 = 2 \cup \{2\} = \left\{\emptyset, \{\emptyset\}, \{\emptyset, \{\emptyset\}\}\right\}$$

and, in general,

$$k' = k \cup \{k\}.$$

It is straightforward to verify Peano's axioms for this model of the natural numbers. Let us first notice that, in this model, the successor n' of a natural number n is given by $n' = n \cup \{n\}$. Now let us sketch the verification of the axioms.

P1 is clear by construction, and so is **P2**.

P3 is an amusing exercise in logic: If $m' = 1$, then there is a set A such that $A \cup \{A\} = 1$ or $A \cup \{A\} = \{\emptyset\}$. In particular, $x \in A \cup \{A\}$ implies $x \in \{\emptyset\}$. Since $A \in A \cup \{A\}$, it follows that $A = \emptyset$. But $\emptyset$ is not a natural number. So 1 is not the successor of a natural number.

For **P4**, it is convenient to invoke the concept of ordering in our model of the natural numbers. Say that $m < n$ if $m \in n$. Clearly, if $m, n \in \mathbb{N}$ and $m \neq n$, then either $m \in n$ or $n \in m$ but not both. Thus we have the usual trichotomy of a strict, simple order. Now suppose that m, n are natural numbers and that $m' = n'$. If $m < n$, then $m \in n$ so $m' \in n'$ and $m' < n'$. That is false. Likewise we cannot have $n < m$. Thus it must be that $m = n$.

We shall discuss **P5** a bit later.

Next we turn to well ordering. We assert that our model of the natural numbers, with the ordering defined in our discussion of **P4**, has the property that if $\emptyset \neq S \subset \mathbb{N}$, then there is an element $s \in S$ such that $s < t$ for every $s \neq t \in S$. It is clear from the trichotomy that the least element s, if it exists, is unique. We proceed in several steps:

Fix a natural number $m > 1$ and restrict attention to $\mathcal{Q}(m)$, which is the set of natural numbers that are less than m.

Proposition 5.1.3 *The set $\mathcal{Q}(m)$ has finitely many elements.*

Proof: A natural number k is less than m if and only if $k \in m$. But, by construction, m has only finitely many elements.   □

Remark 5.1.4 As an exercise, you may wish to attempt to prove this last proposition directly from the Peano axioms.

Proposition 5.1.5 *For each* m, *the set* $\mathcal{Q}(m)$ *is well ordered.*

Proof: The proof is by mathematical induction on m.

When $m = 1$, there is nothing to prove.

Assume that the assertion has been proved for $m = j$. Now let U be a subset of $\mathcal{Q}(j')$. There are now three possibilities:

(1) If in fact $U \subset \mathcal{Q}(j)$, then U has a least element by the inductive hypothesis.

(2) If $U = \{j\}$, then U has but one element and that element, namely j, is the minimum that we seek.

(3) The last possibility is that U contains j and some other natural numbers as well. But then $U \setminus \{j\} \subset \mathcal{Q}(j)$. Hence $U \setminus \{j\}$ has a least element s by the inductive hypothesis. Since s is automatically less than j, it follows that s is a least element for the entire set U.   □

Now we have all our tools in place, and we can prove the full result:

Theorem 5.1.6 *The natural numbers* $\mathbb{N}$ *are well ordered.*

Proof: Let $\emptyset \neq S \subset \mathbb{N}$. Select an element $m \in S$. There are now two possibilities:

(1) If $\mathcal{Q}(m) \cap S = \emptyset$, then m is the least element of S that we seek.

(2) If $T = \mathcal{Q}(m) \cap S \neq \emptyset$, then notice that, if $x \in T$ and $y \in S \setminus T$, then $x < y$. So it suffices for us to find a least element of T. But such an element exists by the preceding proposition.

The proof is complete.   □

We next observe that, in a certain sense, the well-ordering property implies the mathematical induction property (Axiom **P5**). By this we mean the following: Do not consider any model of the natural numbers, but just consider any number system X satisfying **P1** - **P4**. Assume that every element of X except 1 has a predecessor, and in addition assume that this number system is well ordered.

Now let P be a property. Assume that $P(1)$ is true, and assume the syllogism $P(j) \Rightarrow P(j')$ for all j. We claim that $P(m)$ is true for every m. Suppose not. Then the set

$$S = \{m \in X : P(m) \text{ is false}\}$$

is non-empty. Let q be the least element of S; this number is guaranteed to exist by the well-ordering property. The number q cannot be 1, for we assumed that $P(1)$ is true. But if $q > 1$, then q has a predecessor r. Since q is the least element of S, then it cannot be that $r \in S$. Thus $P(r)$ must be true. But then, by our hypothesis, $P(r')$ must be true. However, $r' = q$. So $P(q)$ is true. That is a contradiction.

The only possible conclusion is that S is empty. So $P(n)$ is true for every n.

These remarks about well-ordering implying **P5** are not satisfactory from our point of view, because *we used the inductive property* to establish that every natural number other than 1 has a predecessor. But it is important for you to understand that any development of the natural numbers will result in mathematical induction and well ordering being closely linked.

If you review your calculus book or other elementary texts, you will find that both mathematical induction and well ordering are occasionally used. But in every instance the author will say "These properties of the natural numbers are intuitively clear; trust me." Now you can begin to understand why an elementary textbook author must make that choice. The truth about these topics is inexorably linked to the very foundations of mathematics, and is therefore both subtle and complicated.

The remaining big idea connected with the natural numbers is addition. It can be proven that a satisfactory theory of addition *cannot* be developed from **P1 - P5** (see [SUP, p. 136] for details). Instead, it is customary to adjoin two new axioms to our theory:

P6 If x is a natural number, then $x + 1 = x'$;

P7 For any natural numbers x, y we have

$$x + y' = (x + y)'.$$

To illustrate these ideas, let us close the section by proving that $2 + 2 = 4$. [Judge for yourself whether the proof is as obvious as $2 + 2 = 4$!]. In this argument we use the definitions $2 = 1'$, $3 = 2'$, $4 = 3'$.

$$
\begin{aligned}
2 + 2 &= 2 + 1' && \text{by definition of 2} \\
&= (2 + 1)' && \text{by } \textbf{P7} \\
&= (2')' && \text{by } \textbf{P6} \\
&= 3' && \text{by definition of 3} \\
&= 4 && \text{by definition of 4}
\end{aligned}
$$

In fact, it is even trickier to get multiplication to work in the natural number system. Almost the only viable method is to add even more axioms that control this binary operation. We can say no more about the matter here, but refer to [SUP].

It is not difficult to see that, with enough patience (or with mathematical induction), one could establish all the basic laws of arithmetic. Of course this would not be a fruitful use of our time. The celebrated work [WHB] treats this matter in complete detail.

In the succeeding sections of the present book, we shall take the basic laws of arithmetic on the natural numbers as given. We understand that our treatment of the natural numbers is incomplete. We have touched on some topics, and indicated some constructions. But, when it comes right down to it, we are taking the natural numbers on faith. All of our future number systems (the integers, the rational numbers, the reals, the complexes, the quaternions) will be constructed rigorously. The somewhat bewildering situation before us is that the more complicated numbers systems are easier to construct.

5.2   The Integers

Now we will apply the notion of an equivalence class to *construct* the integers (positive and negative and zero). There is an important point of knowledge to be noted here. In view of Section 5.1, we may take the natural numbers as given. The natural numbers are universally accepted, and we have indicated how they may be constructed in a formal manner. However, the number zero and the negative numbers are a different matter. It was not until the fifteenth century that the concepts of zero and negative numbers started to take hold— for they do not correspond to explicit collections of objects (five fingers or ten shoes) but rather to *concepts* (zero books is the lack of books; minus 4 pens means that we owe someone four pens). After some practice we get used to negative numbers, but explaining in words what they mean is always a bit clumsy.

In fact, it is sobering to realize that the Italian mathematicians of the fifteenth and sixteenth centuries referred to negative numbers—in their *formal writings*—as "fictitious" or "absurd." Mathematics is, in part, a subject that we must get used to. It took several hundred years for mankind to get used to negative numbers.

It is much more satisfying, from the point of view of logic, to *construct* the integers from what we already have, that is, from the natural numbers. We proceed as follows. Let $A = \mathbb{N} \times \mathbb{N}$, the set of ordered pairs of natural numbers. We define a relation $\mathcal{R}$ on A as follows:

$$(a, b) \text{ is related to } (a^*, b^*) \text{ if } a + b^* = a^* + b$$

Theorem 5.2.1 *The relation $\mathcal{R}$ is an equivalence relation.*

Proof: That (a, b) is related to (a, b) follows from the trivial identity $a + b = a + b$. Hence $\mathcal{R}$ is reflexive. Second, if (a, b) is related to (a^*, b^*), then

$a + b^* = a^* + b$ hence $a^* + b = a + b^*$ (just reverse the equality) hence (a^*, b^*) is related to (a, b). So $\mathcal{R}$ is symmetric.

Finally, if (a, b) is related to (a^*, b^*) and (a^*, b^*) is related to (a^{**}, b^{**}), then we have

$$a + b^* = a^* + b \qquad \text{and} \qquad a^* + b^{**} = a^{**} + b^*.$$

Adding these equations gives

$$(a + b^*) + (a^* + b^{**}) = (a^* + b) + (a^{**} + b^*).$$

Cancelling a^* and b^* from each side finally yields

$$a + b^{**} = a^{**} + b.$$

Thus (a, b) is related to (a^{**}, b^{**}). Therefore $\mathcal{R}$ is transitive. We conclude that $\mathcal{R}$ is an equivalence relation. $\qquad \qquad \square$

Remark 5.2.2 We cheated a bit in the proof of Theorem 5.2.1. Since we do not yet have negative numbers, we therefore have not justified the process of "cancelling" that we used. The most rudimentary form of cancellation is Axiom **P4** of the natural numbers. Suggest a way to use mathematical induction, together with Axiom **P4**, to prove that if a, b, c are natural numbers and if $a + b = c + b$, then $a = c$.

Now our job is to understand the equivalence classes that are induced by $\mathcal{R}$. Let $(a, b) \in A = \mathbb{N} \times \mathbb{N}$, and let $[(a, b)]$ be the corresponding equivalence class. If $b > a$, then we will denote this equivalence class by the integer $b - a$. For instance, the equivalence class $[(2, 7)]$ will be denoted by 5. Notice that if $(a^*, b^*) \in [(a, b)]$, then $a + b^* = a^* + b$ hence $b^* - a^* = b - a$ *as long as $b > a$*. Therefore the numeral that we choose to represent our equivalence class is *independent of which element of the equivalence class is used to compute it*.

If $(a, b) \in A$ and $b = a$, then we let the symbol 0 denote the equivalence class $[(a, b)]$. Notice that if (a^*, b^*) is any other element of this particular $[(a, b)]$, then it must be that $a + b^* = a^* + b$ hence $b^* = a^*$; therefore this definition is unambiguous.

If $(a, b) \in A$ and $a > b$, then we will denote the equivalence class $[(a, b)]$ by the symbol $-(a - b)$. For instance, we will denote the equivalence class $[(7, 5)]$ by the symbol -2. Once again, if (a^*, b^*) is related to (a, b), then the equation $a + b^* = a^* + b$ guarantees that our choice of symbol to represent $[(a, b)]$ is unambiguous.

Thus we have given our equivalence classes names, and these names *look just like* the names that we give to integers: there are positive integers, and negative ones, and zero. But we want to see that these objects *behave* like integers. [As you read on, use the informal mnemonic that the equivalence class $[(a, b)]$ stands for the integer $b - a$.]

First, do these new objects that we have constructed *add* correctly? Well, let $A = [(a, b)]$ and $C = [(c, d)]$ be two equivalence classes. *Define* their sum to be $A + C = [(a + c, b + d)]$. We must check that this is unambiguous. If $(\tilde{a}, \tilde{b})$ is related to (a, b) and $(\tilde{c}, \tilde{d})$ is related to (c, d), then of course we know that

$$a + \tilde{b} = \tilde{a} + b$$

and

$$c + \tilde{d} = \tilde{c} + d.$$

Adding these two equations gives

$$(a + c) + (\tilde{b} + \tilde{d}) = (\tilde{a} + \tilde{c}) + (b + d)$$

hence $(a+c, b+d)$ is related to $(\tilde{a}+\tilde{c}, \tilde{b}+\tilde{d})$. Thus adding two of our equivalence classes gives another equivalence class, as it should. We say that addition of integers is *well defined.*

This point is so significant that it bears repeating. Each integer is an equivalence class—that is, a *set.* If we are going to add two integers m and n by choosing an element from the set m and another element from the set n, then the operation that we define had better be independent of the choice of elements. This is another way of saying that we want the sum of two equivalence classes to be another equivalence class. We call this the concept of "well definedness."

EXAMPLE 5.2.3 To add 5 and 3, we first note that 5 is the equivalence class $[(2, 7)]$ and 3 is the equivalence class $[(2, 5)]$. We add them componentwise and find that the sum is

$$[(2, 7)] + [2, 5)] = [(2 + 2, 7 + 5)] = [(4, 12)].$$

Which equivalence class is this answer? Looking back at our prescription for giving names to the equivalence classes, we see that this is the equivalence class that we called $12 - 4$ or 8. So we have rediscovered the fact that $5 + 3 = 8$.

Now let us add 4 and -9. The first of these is the equivalence class $[(3, 7)]$, and the second is the equivalence class $[(13, 4)]$. The sum is therefore $[(16, 11)]$, and this is the equivalence class that we call $-(16 - 11)$ or -5. That is the answer that we would expect when we add 4 to -9.

Next, we add -12 and -5. Previous experience leads us to expect the answer to be -17. Now -12 is the equivalence class $[(19, 7)]$, and -5 is the equivalence class $[(7, 2)]$. The sum is $[(26, 9)]$, which is the equivalence class that we call -17.

Finally, we can see in practice that our method of addition is unambiguous. Let us redo the second example using $[(6, 10)]$ as the equivalence class denoted by 4 and $[(15, 6)]$ as the equivalence class denoted by -9. Then the sum is $[(21, 16)]$, and this is still the equivalence class -5, as it should be.   ◇

Remark 5.2.4 What is the point of this section? Everyone knows about negative numbers, so why go through this abstract construction? The reason is that, until one sees this construction, negative numbers are just imaginary objects—placeholders if you will—which are a useful notation but which do not exist. Now they *do* exist. They are a collection of equivalence classes of pairs of natural numbers. This collection is equipped with certain arithmetic operations, such as addition, subtraction, and multiplication. We now discuss these last two.

If $A = [(a, b)]$ and $C = [(c, d)]$ are integers, then we define their *difference* to be the equivalence class $[(a + d, b + c)]$; we denote this difference by $A - C$. [Note that we may not use subtraction of natural numbers in our definition of subtraction of integers; subtraction of natural numbers is not, in general, defined.] The unambiguity (or well definedness) of this definition is treated in the exercises.

EXAMPLE 5.2.5 We calculate $8 - 14$. Now $8 = [(1, 9)]$ and $14 = [(3, 17)]$. Therefore
$$8 - 14 = [(1 + 17, 9 + 3)] = [(18, 12)] = -6,$$
as expected.

As a second example, we compute $(-4) - (-8)$. Now
$$-4 - (-8) = [(6, 2)] - [(13, 5)] = [(6 + 5, 2 + 13)] = [(11, 15)] = 4.$$

◇

Remark 5.2.6 When we first learn that $(-4) - (-8) = (-4) + 8 = 4$, the explanation is a bit mysterious: why is "minus a minus equal to a plus"? Now there is no longer any mystery: this property follows *from our construction* of the number system $\mathbb{Z}$.

Remark 5.2.7 It is interesting to sort out the last example from the justification for the arithmetic of negative numbers that we learn in high school. Here is an example of that reasoning.

It is postulated that negative numbers exist (they certainly are not constructed). Then it is noted that
$$18 + (8 - 14) = (18 - 14) + 8 = 4 + 8 = 12 = 18 - 6 = 18 + (-6).$$

Identifying the far left and far right sides of the equation, we cancel 18 from each side and conclude that $8 - 14 = -6$.

This reasoning is perfectly correct. But it presupposes the existence of a number system that (**i**) contains negative integers and (**ii**) obeys all the familiar laws of arithmetic.

The advantage of the presentation in this section of the present book is that we actually *construct* such a number system. We do not presuppose it. The additive properties of negative numbers follow automatically from our construction. They are not derived by algebraic tricks from some numbers that we do not actually know exist.

Finally, we turn to multiplication. If $A = [(a, b)]$ and $C = [(c, d)]$ are integers, then we define their product by the formula

$$A \cdot C = [(a \cdot d + b \cdot c, a \cdot c + b \cdot d)].$$

This definition may be a surprise. Why did we not define $A \cdot C$ to be $[(a \cdot c, b \cdot d)]$? There are several reasons: first of all, the latter definition would give the wrong answer; moreover, it is not unambiguous (different representatives of A and C would give a different answer). If you recall that we think of $[(a, b)]$ as representing $b - a$ and $[(c, d)]$ as representing $d - c$, then the product should be the equivalence class that represents $(b - a) \cdot (d - c)$. That is the motivation behind our definition.

The unambiguity (or well definedness) of the given definition of multiplication of integers is treated in the exercises. We proceed now to an example.

EXAMPLE 5.2.8 We compute the product of -3 and -6. Now

$$(-3) \cdot (-6) = [(5, 2)] \cdot [(9, 3)] = [(5 \cdot 3 + 2 \cdot 9, 5 \cdot 9 + 2 \cdot 3)] = [(33, 51)] = 18,$$

which is the expected answer.

As a second example, we multiply -5 and 12. We have

$$-5 \cdot 12 = [(7, 2)] \cdot [(1, 13)] = [(7 \cdot 13 + 2 \cdot 1, 7 \cdot 1 + 2 \cdot 13)] = [(93, 33)] = -60.$$

Finally, we show that 0 times any integer A equals 0. Let $A = [(a, b)]$. And let $0 = [(1, 1)]$. Then

$$0 \cdot A = [(1, 1)] \cdot [(a, b)] = [(1 \cdot b + 1 \cdot a, 1 \cdot a + 1 \cdot b)] = [(a + b, a + b)] = 0.$$

Remark 5.2.9 Notice that one of the pleasant by-products of our construction of the integers is that we no longer have to give artificial explanations for why the product of two negative numbers is a positive number or why the product of a negative number and a positive number is negative. These properties instead follow *automatically* from our construction.

Remark 5.2.10 It is interesting to sort out the last example from the justification for the arithmetic of negative numbers that we learn in high school. Here is an example of that reasoning.

It is postulated that negative numbers exist (they certainly are not constructed). Then it is noted that

$$3 \cdot 8 = (6 - 3) \cdot 8 = 6 \cdot 8 - 3 \cdot 8$$

hence

$$24 = 48 - 3 \cdot 8$$

or, using reasoning as in our last remark but one,

$$-24 = -3 \cdot 8.$$

Similarly, one can show that

$$-48 = -6 \cdot 8\,.$$

Taking these two facts for granted, we then compute that

$$(8 - 3) \cdot (8 - 6) = 8 \cdot 8 + 8 \cdot (-6) + (-3) \cdot 8 + (-3) \cdot (-6)\,.$$

As a result,

$$10 = 64 - 48 - 24 + (-3) \cdot (-6)$$

or

$$10 + 72 - 64 = (-3) \cdot (-6)\,,$$

hence

$$18 = (-3) \cdot (-6)\,.$$

Again, this reasoning is perfectly correct. But it presupposes the existence of a number system that **(i)** contains negative integers and **(ii)** obeys all the familiar laws of arithmetic.

The advantage of the presentation in this section of the present book is that we actually *construct* such a number system. We do not presuppose it. The multiplicative properties of negative numbers follow automatically from our construction. They are not derived by algebraic tricks from some numbers that we do not actually know exist.

Notice that the integers $\mathbb{Z}$ as we have constructed them contain the element $0 \equiv [(1,1)]$. This element is the *additive identity* in the sense that $x + 0 = 0 + x = x$ for any integer x. Also, if $y = [(a,b)]$ is any integer, then it has an *additive inverse* $-y = [(b,a)]$. This means that $y + (-y) = 0$. As a result of these two facts, the integers $\mathbb{Z}$ form a *group*. We shall say more about groups in Section 9.1.

The integer system that we have constructed also contains a multiplicative identity. It is $1 = [(0,1)]$. Check for yourself that, if $A = [(a,b)]$ is any integer, then $1 \cdot A = A \cdot 1 = A$.

Of course we will not discuss division for integers; in general, division of one integer by another makes no sense *in the universe of the integers*. More will be said about this fact in the exercises.

In the rest of this book, we shall follow the standard mathematical custom of denoting the set of all integers by the symbol $\mathbb{Z}$. We will write the integers not as equivalence classes, but in the usual way as the sequence of digits $\cdots - 3, -2, -1, 0, 1, 2, 3, \ldots$. The equivalence classes are a device that we used to *construct* the integers. Now that we have them, we may as well write them in the simple, familiar fashion and manipulate them as usual.

In an exhaustive treatment of the construction of $\mathbb{Z}$, we would prove that addition and multiplication are commutative and associative, prove the distributive law, and so forth. But the purpose of this section is to demonstrate modes of logical thought rather than to be exhaustive. We shall say more about some of the elementary properties of the integers in the exercises.

5.3   The Rational Numbers

In this section we use the integers, together with a construction using equiv-
alence classes, to build the rational numbers. Let A be the set $\mathbb{Z} \times (\mathbb{Z} \setminus \{0\})$.
In other words, A is the set of ordered pairs (a, b) of integers subject to the
condition that $b \neq 0$. [*Think of this ordered pair as ultimately "representing"
the fraction a/b.*] We definitely want it to be the case that certain ordered
pairs represent the same number. For instance,

$$\tfrac{1}{2} \text{ should be the same number as } \tfrac{3}{6}.$$

This motivates our equivalence relation. Declare (a, b) to be related to (a^*, b^*)
if $a \cdot b^* = a^* \cdot b$. [*Here we are thinking that the fraction a/b should equal the
fraction a*/b* precisely when $a \cdot b^* = a^* \cdot b$.*]

Is this an equivalence relation? Obviously the pair (a, b) is related to itself,
since $a \cdot b = a \cdot b$. Also the relation is symmetric: if (a, b) and (a^*, b^*) are pairs
and $a \cdot b^* = a^* \cdot b$, then $a^* \cdot b = a \cdot b^*$. Finally, if (a, b) is related to (a^*, b^*) and
(a^*, b^*) is related to (a^{**}, b^{**}), then we have both

$$a \cdot b^* = a^* \cdot b \text{ and } a^* \cdot b^{**} = a^{**} \cdot b^*. \tag{$\star$}$$

Multiplying the left sides of these two equations together and the right sides
together gives

$$(a \cdot b^*) \cdot (a^* \cdot b^{**}) = (a^* \cdot b) \cdot (a^{**} \cdot b^*). \tag{$\star\star$}$$

If $a^* = 0$, then it follows immediately from $(\star)$ that both a and a^{**} must
be zero. So the three pairs $(a, b), (a^*, b^*)$, and (a^{**}, b^{**}) are equivalent, and
there is nothing to prove. So we may assume that $a^* \neq 0$. We know *a priori*
that $b^* \neq 0$; therefore we may cancel common terms in the equation $(\star\star)$ to
obtain

$$a \cdot b^{**} = b \cdot a^{**}.$$

Thus (a, b) is related to (a^{**}, b^{**}), and our relation is transitive. [Exercise:
explain why it is correct to "cancel common terms" in the last step.]

The resulting collection of equivalence classes will be called the set of
rational numbers, and we shall denote this set with the symbol $\mathbb{Q}$.

EXAMPLE 5.3.1 The equivalence class $[(4, 12)]$ contains all of the pairs
$(4, 12), (1, 3), (-2, -6)$. (Of course it contains infinitely many other pairs as
well.) This equivalence class represents the fraction $4/12$ which we sometimes
also write as $1/3$ or $(-2)/(-6)$.   ◇

If $[(a, b)]$ and $[(c, d)]$ are rational numbers then we define their *product* to
be the rational number

$$[(a \cdot c, b \cdot d)].$$

This is well defined (unambiguous), for the following reason. Suppose that (a, b) is related to $(\tilde{a}, \tilde{b})$ and (c, d) is related to $(\tilde{c}, \tilde{d})$. We would like to know that $[(a, b)] \cdot [(c, d)] = [(a \cdot c, b \cdot d)]$ is the same equivalence class as $[(\tilde{a}, \tilde{b})] \cdot [(\tilde{c}, \tilde{d})] = [(\tilde{a} \cdot \tilde{c}, \tilde{b} \cdot \tilde{d})]$. In other words, we need to know that

$$(a \cdot c) \cdot (\tilde{b} \cdot \tilde{d}) = (\tilde{a} \cdot \tilde{c}) \cdot (b \cdot d). \tag{$*$}$$

But our hypothesis is that

$$a \cdot \tilde{b} = \tilde{a} \cdot b \quad \text{and} \quad c \cdot \tilde{d} = \tilde{c} \cdot d.$$

Multiplying together the left sides and the right sides, we obtain

$$(a \cdot \tilde{b}) \cdot (c \cdot \tilde{d}) = (\tilde{a} \cdot b) \cdot (\tilde{c} \cdot d).$$

Rearranging, we have

$$(a \cdot c) \cdot (\tilde{b} \cdot \tilde{d}) = (\tilde{a} \cdot \tilde{c}) \cdot (b \cdot d).$$

But this is just $(*)$. So multiplication is unambiguous or well defined.

EXAMPLE 5.3.2 The product of the two rational numbers $[(3, 8)]$ and $[(-2, 5)]$ is

$$[(3 \cdot (-2), 8 \cdot 5)] = [(-6, 40)] = [(-3, 20)].$$

This is what we expect: the product of 3/8 and $-2/5$ is $-3/20$. ◇

If $q = [(a, b)]$ and $r = [(c, d)]$ are rational numbers and if r is not zero (that is, $[(c, d)]$ is not the equivalence class zero—in other words, $c \neq 0$), then we define the quotient q/r to be the equivalence class

$$[(ad, bc)].$$

We leave it to you to check that this operation is well defined.

EXAMPLE 5.3.3 The quotient of the rational number $[(4, 7)]$ by the rational number $[(3, -2)]$ is, by definition, the rational number

$$[(4 \cdot (-2), 7 \cdot 3)] = [(-8, 21)].$$

This is what we expect: the quotient of 4/7 by $-3/2$ is $-8/21$. ◇

How should we add two rational numbers? We could try declaring $[(a, b)] + [(c, d)]$ to be $[(a + c, b + d)]$, but this will not work (think about the way that we usually add fractions). Instead we define

$$[(a, b)] + [(c, d)] = [(a \cdot d + b \cdot c, b \cdot d)].$$

That this definition is well defined (unambiguous) is left for the exercises. We turn instead to an example.

EXAMPLE 5.3.4 The sum of the rational numbers $[(3, -14)]$ and $[(9, 4)]$ is given by

$$[(3 \cdot 4 + (-14) \cdot 9, (-14) \cdot 4)] = [(-114, -56)] = [(57, 28)].$$

This coincides with the usual way that we add fractions :

$$-\frac{3}{14} + \frac{9}{4} = \frac{57}{28}.$$

$\diamond$

Notice that the equivalence class $[(0, 1)]$ is the rational number that we usually denote by 0. It is the additive identity, for if $[(a, b)]$ is another rational number, then

$$[(0, 1)] + [(a, b)] = [(0 \cdot b + 1 \cdot a, 1 \cdot b)] = [(a, b)].$$

A similar argument shows that $[(0, 1)]$ times any rational number gives $[(0, 1)]$ or 0. By the same token, the rational number $[(1, 1)]$ is the multiplicative identity. We leave the details for you.

Of course the concept of subtraction is really just a special case of addition (that is $\alpha - \beta$ is the same thing as $\alpha + (-\beta)$). So we shall say nothing further about subtraction.

In practice we will write rational numbers in the traditional fashion:

$$\frac{2}{5}, \quad \frac{-19}{3}, \quad \frac{22}{2}, \quad \frac{24}{4}, \quad \ldots .$$

In mathematics it is generally not wise to write rational numbers in mixed form, such as $2\frac{3}{5}$, because the juxtaposition of two numbers could easily be mistaken for multiplication. Instead, we would write this quantity as the improper fraction $13/5$.

Definition 5.3.5 A set S is called a *field* if it is equipped with a binary operation (usually called addition and denoted by "+") and a second binary operation (usually called multiplication and denoted by "·") such that the following axioms are satisfied:

A1. S is closed under addition: if $x, y \in S$, then $x + y \in S$.
A2. Addition is commutative: if $x, y \in S$, then $x + y = y + x$.
A3. Addition is associative: if $x, y, z \in S$, then
$$x + (y + z) = (x + y) + z.$$
A4. There exists an element, called 0, in S which is an additive identity: if $x \in S$, then $0 + x = x$.
A5. Each element of S has an additive inverse: if $x \in S$ then there is an element $-x \in S$ such that $x + (-x) = 0$.

M1. S is closed under multiplication: if $x, y \in S$, then $x \cdot y \in S$.
M2. Multiplication is commutative: if $x, y \in S$, then $x \cdot y = y \cdot x$.

M3. Multiplication is associative: if $x, y, z \in S$, then
$$x \cdot (y \cdot z) = (x \cdot y) \cdot z.$$
M4. There exists an element, called 1, which is a multiplicative identity: if $x \in S$, then $1 \cdot x = x$.
M5. Each non-zero element of S has a multiplicative inverse: if $0 \neq x \in S$, then there is an element $x^{-1} \in S$ such that $x \cdot (x^{-1}) = 1$. The element x^{-1} is sometimes denoted by $1/x$.
D1. Multiplication distributes over addition: if $x, y, z \in S$, then
$$x \cdot (y + z) = x \cdot y + x \cdot z.$$

Eleven axioms is a lot to digest all at once, but in fact these are all familiar properties of addition and multiplication of rational numbers that we use every day: the set $\mathbb{Q}$, with the usual notions of addition and multiplication (and with the usual additive identity 0 and multiplicative identity 1), forms a field. The integers, by contrast, do not: non-zero elements of $\mathbb{Z}$ (except 1 and -1) do not have multiplicative inverses *in the integers*.

Let us now consider some consequences of the field axioms.

Theorem 5.3.6 *Any field has the following properties:*

(1) *If $z + x = z + y$, then $x = y$.*

(2) *If $x + z = 0$, then $z = -x$ (the additive inverse is unique).*

(3) $-(-y) = y$.

(4) *If $y \neq 0$ and $y \cdot x = y \cdot z$, then $x = z$.*

(5) *If $y \neq 0$ and $y \cdot z = 1$, then $z = y^{-1}$ (the multiplicative inverse is unique).*

(6) $\left(x^{-1}\right)^{-1} = x$.

(7) $0 \cdot x = 0$.

(8) *If $x \cdot y = 0$, then either $x = 0$ or $y = 0$.*

(9) $(-x) \cdot y = -(x \cdot y) = x \cdot (-y)$.

(10) $(-x) \cdot (-y) = x \cdot y$.

Proof: These are all familiar properties of the rationals, but now we are considering them for an arbitrary field. We prove just a few to illustrate the logic. The proofs of the others are assigned as exercises.

To prove **(1)**, we write
$$z + x = z + y \Rightarrow (-z) + (z + x) = (-z) + (z + y)$$

and now Axiom A3 yields that this implies
$$((-z) + z) + x = ((-z) + z) + y.$$

Next, Axiom A5 yields that

$$0 + x = 0 + y$$

and hence, by Axiom A4,

$$x = y.$$

To prove **(7)**, we observe that

$$0 \cdot x = (0 + 0) \cdot x.$$

By Axiom M2, the right-hand side equals

$$x \cdot (0 + 0).$$

By Axiom D1, the last expression equals

$$x \cdot 0 + x \cdot 0,$$

which by Axiom M2 equals $0 \cdot x + 0 \cdot x$. Thus we have derived the equation

$$0 \cdot x = 0 \cdot x + 0 \cdot x.$$

Axioms A4 and A2 let us rewrite the left side as

$$0 \cdot x + 0 = 0 \cdot x + 0 \cdot x.$$

Finally, part **(1)** of the present theorem (which we have already proved) yields that

$$0 = 0 \cdot x,$$

which is the desired result.

To prove **(8)**, we suppose that $x \neq 0$. In this case, x has a multiplicative inverse x^{-1} and we multiply both sides of our equation by this element:

$$x^{-1} \cdot (x \cdot y) = x^{-1} \cdot 0.$$

By Axiom M3, the left side can be rewritten and we have

$$(x \cdot x^{-1}) \cdot y = x^{-1} \cdot 0.$$

Next, we rewrite the right side using Axiom M2:

$$(x \cdot x^{-1}) \cdot y = 0 \cdot x^{-1}.$$

Now Axiom M5 allows us to simplify the left side:

$$1 \cdot y = 0 \cdot x^{-1}.$$

We further simplify the left side using Axiom M4 and the right side using Part **(7)** of the present theorem (which we just proved) to obtain:

$$y = 0.$$

Thus we see that if $x \neq 0$ then $y = 0$. But this is logically equivalent with $x = 0$ or $y = 0$, as we wished to prove. [If you have forgotten why these statements are logically equivalent, write a truth table.] □

Refer to Section 4.2 for the notion of a strict, simple ordering.

EXAMPLE 5.3.7 The integers $\mathbb{Z}$ form a strictly, simply ordered set when equipped with the usual ordering. We can make this ordering precise by saying that $x < y$ if $y - x$ is a positive integer. For instance,

$$6 < 8 \quad \text{because} \quad 8 - 6 = 2 > 0.$$

Likewise,

$$-5 < -1 \quad \text{because} \quad -1 - (-5) = 4 > 0.$$

Observe that the same ordering works on the rational numbers. ◇

If A is a strictly ordered set and a, b are elements, then we often write $a \leq b$ to mean that *either $a < b$ or $a = b$.*

When a field has an ordering which is compatible with the field operations, then a richer structure results:

Definition 5.3.8 A field F is called an *ordered field* if F has a strict, simple ordering $<$ that satisfies the following addition properties:

(1) If $x, y, z \in F$ and if $y < z$, then $x + y < x + z$.
(2) If $x, y \in F, x > 0$, and $y > 0$, then $x \cdot y > 0$.

Again, these are familiar properties of the rational numbers: $\mathbb{Q}$ forms an ordered field. Some further properties of ordered fields may be proved from the axioms:

Theorem 5.3.9 *Any ordered field has the following properties:*

(1) *If $x > 0$ and $z < y$, then $x \cdot z < x \cdot y$.*
(2) *If $x < 0$ and $z < y$, then $x \cdot z > x \cdot y$.*
(3) *If $x > 0$, then $-x < 0$. If $x < 0$, then $-x > 0$.*
(4) *If $0 < y < x$, then $0 < 1/x < 1/y$.*
(5) *If $x \neq 0$, then $x^2 > 0$.*
(6) *If $0 < x < y$, then $x^2 < y^2$.*

Proof: Again we prove just a few of these statements and leave the rest as exercises.

To prove **(1)**, observe that property **(1)** of ordered fields together with our hypothesis implies that

$$(-z) + z < (-z) + y.$$

Thus, using **A2**, we see that $y - z > 0$. Since $x > 0$, property **(2)** of ordered fields gives

$$x \cdot (y - z) > 0.$$

Finally,

$$x \cdot y = x \cdot [(y - z) + z] = x \cdot (y - z) + x \cdot z > 0 + x \cdot z$$

(by property **(1)** of ordered fields again). In conclusion,

$$x \cdot y > x \cdot z.$$

To prove **(3)**, begin with the equation

$$0 = -x + x.$$

Since $x > 0$, the right side is greater than $-x$. Thus $0 > -x$ as claimed. The proof of the other statement of **(3)** is similar.

To prove **(5)**, we consider two cases. If $x > 0$, then $x^2 \equiv x \cdot x$ is positive by property **(2)** of ordered fields. If $x < 0$, then $-x > 0$ (by part **(3)** of the present theorem, which we just proved) hence $(-x) \cdot (-x) > 0$. But part **(10)** of the last theorem guarantees that $(-x) \cdot (-x) = x \cdot x$ hence $x \cdot x > 0$. $\square$

5.4   The Real Number System

Now that we are accustomed to the notion of equivalence classes, the construction of the integers and of the rational numbers seems fairly natural. In fact, equivalence classes provide a precise language for declaring certain objects to be equal (or for identifying certain objects). We can now use the integers and the rationals as we always have done, with the added confidence that they are not simply a useful notation but that they have been *constructed*.

We turn next to the real numbers. We saw in Section 2.3 that the rational number system is not closed under the operation of taking square roots, for example. We know from calculus that for many other purposes the rational numbers are inadequate. It is important to work in a number system that is closed with respect to all the operations we shall perform. While the rationals are closed under the usual arithmetic operations, they are not closed under the operation of taking *limits*. For instance, the sequence of rational numbers $3, 3.1, 3.14, 3.141, 3.1415, 3.14159, \ldots$ consists of terms that seem to be getting closer and closer together, *seem* to tend to some limit, and yet there is no rational number which will serve as a limit (of course it turns out that the limit is π—an "irrational" number).

We will now deal with the real number system, a system that contains all limits of sequences of rational numbers (as well as all limits of sequences of real numbers!). In fact, our plan will be as follows: in this section we shall discuss all the requisite properties of the reals. The actual construction of the reals is rather complicated, and we shall put that in the optional next section.

Definition 5.4.1 Let A be an ordered set and X a subset of A. The set X is called *bounded above* if there is an element $b \in A$ such that $x \leq b$ for all $x \in X$. We call the element b an *upper bound* for the set X.

EXAMPLE 5.4.2 Let $A = \mathbb{Q}$ with the usual ordering. The set $X = \{x \in \mathbb{Q} : 2 < x < 4\}$ is bounded above. For example, the number 15 is an upper bound for X. So are the numbers 12 and 4. It is interesting to observe that no element of this particular X can be an upper bound for X. The number 4 is a good candidate, but 4 is not an element of X. In fact if $b \in X$ then $(b + 4)/2 \in X$ and $b < (b + 4)/2$, so b could not be an upper bound for X. $\diamond$

It turns out that the most convenient way to formulate the notion that the real numbers have "no gaps" (i.e., that all sequences that seem to be converging actually have something to converge to) is in terms of upper bounds.

Definition 5.4.3 Let A be an ordered set and X a subset of A. An element $b \in A$ is called a *least upper bound* (or *supremum*) for X if b is an upper bound for X and there is no upper bound b^* for X with $b^* < b$. We denote the supremum/(least upper bound) of X by $\sup X$ or $\operatorname{lub} X$.

By its very definition, if a least upper bound exists, then it is unique.

EXAMPLE 5.4.4 In the last example, we considered the set X of rational numbers strictly between 2 and 4. We observed there that 4 is the least upper bound for X. Note that this least upper bound is not an element of the set X.

The set $Y = \{y \in \mathbb{Z} : -9 \leq y \leq 7\}$ has least upper bound 7. In this case, the least upper bound *is* an element of the set Y. $\diamond$

Notice that we may define a lower bound for a subset of an ordered set in a fashion similar to that for an upper bound: $\ell \in A$ is a lower bound for $X \subset A$ if $\ell \leq x$ for all $x \in X$. A *greatest lower bound* (or *infimum*) for X is then defined to be a lower bound ℓ such that there is no lower bound ℓ^* with $\ell^* > \ell$. We denote the infimum/(greatest lower bound) of X by $\inf X$ or $\operatorname{glb} X$.

EXAMPLE 5.4.5 The set X in Examples 6.4.2, 6.4.4 has lower bounds -20, $0, 1, 2$, for instance. The greatest lower bound is 2, which is *not* an element of the set.

The set Y in the last example has lower bounds $-53, -22, -10, -9$, to name just a few. The number -9 is the greatest lower bound. It *is* an element of Y. $\diamond$

EXAMPLE 5.4.6 Let $S = \mathbb{Z} \subset \mathbb{R}$. Then S does not have either an upper bound or a lower bound.   ◇

The purpose that the real numbers will serve for us is as follows: they will contain the rationals, they will still be an ordered field, and *every non-empty subset which has an upper bound will have a least upper bound*. We formulate this property as a theorem.

Theorem 5.4.7 *There exists an ordered field $\mathbb{R}$ that (i) contains $\mathbb{Q}$ and (ii) has the property that any non-empty subset of $\mathbb{R}$ which has an upper bound has a least upper bound.*

The last property described in this theorem is called the Least Upper Bound Property of the real numbers. As mentioned previously, this theorem will be proved in the optional next section. Now we begin to realize why it is so important to *construct* the number systems that we will use. We are endowing $\mathbb{R}$ with a great many properties. Why do we have any right to suppose that there exists a number system with all these properties? We must produce one!

Let us begin to explore the richness of the real numbers. The next theorem states a property that is certainly not shared by the rationals (see Section 2.3). It is fundamental in its importance.

Theorem 5.4.8 *Let x be a positive real number. Then there is a positive real number y such that $y^2 = y \cdot y = x$.*

Proof: We will use throughout this proof the fact (see Part **6** of Theorem 5.3.9) that if $0 < a < b$, then $a^2 < b^2$.

Let

$$S = \{s \in \mathbb{R} : s > 0 \text{ and } s^2 < x\}.$$

Then S is not empty since $x/2 \in S$ if $x < 2$ and $1 \in S$ otherwise. Also S is bounded above since $x + 1$ is an upper bound for S. By Theorem 5.4.7, the set S has a least upper bound. Call it y. Obviously $0 < \min\{x/2, 1\} \le y$ hence y is positive. We claim that $y^2 = x$. To see this, we eliminate the other two possibilities.

If $y^2 < x$, then set $\varepsilon = (x - y^2)/[4(x + 1)]$. Then $\varepsilon > 0$ and

$$
\begin{aligned}
(y + \varepsilon)^2 &= y^2 + 2 \cdot y \cdot \varepsilon + \varepsilon^2 \\
&= y^2 + 2 \cdot y \cdot \frac{x - y^2}{4(x + 1)} + \frac{x - y^2}{4(x + 1)} \cdot \frac{x - y^2}{4(x + 1)} \\
&< y^2 + 2 \cdot \frac{y}{x + 1} \cdot \frac{x - y^2}{4} + \frac{x - y^2}{4} \cdot \frac{x}{4x} \\
&< y^2 + \frac{x - y^2}{2} + \frac{x - y^2}{16} \\
&< y^2 + (x - y^2) \\
&= x.
\end{aligned}
$$

Thus $y + \varepsilon \in S$, and y cannot be an upper bound for S. This contradiction tells us that $y^2 \not< x$.

Similarly, if it were the case that $y^2 > x$, then we set $\varepsilon = (y^2 - x)/[4(x+1)]$. A calculation like the one we just did (see Exercise 5.37) then shows that $(y - \varepsilon)^2 > x$. Hence $y - \varepsilon$ is also an upper bound for S, and y is therefore not the *least* upper bound. This contradiction shows that $y^2 \not> x$.

The only remaining possibility is that $y^2 = x$. That completes the proof. $\square$

A similar proof shows that, if n is a positive integer and x a positive real number, then there is a positive real number y such that $y^n = x$. Exercise 5.38 asks you to provide the details.

We next use the Least Upper Bound Property of the real numbers to establish two important qualitative properties of the real numbers:

Theorem 5.4.9 *The set $\mathbb{R}$ of real numbers satisfies the Archimedean Property:*

Let a and b be positive real numbers. Then there is a natural number n such that $na > b$.

Theorem 5.4.10 *The set $\mathbb{Q}$ of rational numbers satisfies the following Density Property:*

Let $c < d$ be real numbers. Then there is a rational number q with $c < q < d$.

Proof of Theorem 5.4.9: Suppose the Archimedean Property to be false. Then $S = \{na : n \in \mathbb{N}\}$ has b as an upper bound. Therefore S has a finite supremum β. Since $a > 0, \beta - a < \beta$. So $\beta - a$ is not an upper bound for S, and there must be a natural number n^* such that $n^* \cdot a > \beta - a$. But then $(n^* + 1)a > \beta$, and β cannot be the supremum for S. This contradiction proves the theorem. $\square$

Proof of Theorem 5.4.10: Let $\lambda = d - c > 0$. By the Archimedean Property, choose a positive integer N such that $N \cdot \lambda > 1$. Again, the Archimedean Property gives a natural number P such that $P > N \cdot c$ and another Q such that $Q > | - N \cdot c|$. Then $Q > -N \cdot c$, and we see that Nc falls between the integers $-Q$ and P; therefore there must be an integer M between $-Q$ and P (inclusive) such that
$$M - 1 \leq Nc < M.$$
Thus $c < M/N$. Also
$$M \leq Nc + 1 \quad \text{hence} \quad \frac{M}{N} \leq c + \frac{1}{N} < c + \lambda = d.$$

So M/N is a rational number lying strictly between c and d. $\square$

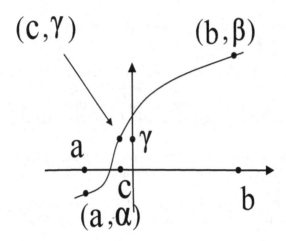

FIGURE 5.1
The Intermediate Value Property.

One of the most profound and useful properties of the real numbers, and one that is equivalent to the Least Upper Bound Property, is the Intermediate Value Property:

Theorem 5.4.11 *Let f be a continuous, real-valued function with domain the interval $[a, b]$. If $f(a) = \alpha$, $f(b) = \beta$, and if $\alpha < \gamma < \beta$, then there is a value $c \in (a, b)$ such that $f(c) = \gamma$. Refer to Figure 5.1.*

Proof: Let
$$S = \{x \in [a, b] : f(x) < \gamma\}.$$
Then $S \neq \emptyset$ since $a \in S$. Moreover, S is bounded above by b. So $c = \sup S$ exists as a finite real number. We claim that $f(c) = \gamma$.

Clearly $f(c) \leq \gamma$ since c is the limit of numbers at which f takes values less than γ (we use the continuity of f here). Suppose, seeking a contradiction, that $f(c) < \gamma$. Let $\epsilon = \gamma - f(c)$. By the continuity of f, we may select $\delta > 0$ such that $|t - c| < \delta$ implies that $|f(t) - f(c)| < \epsilon/2$. But then, for $t \in (c - \delta, c + \delta)$, $f(t) < f(c) + \epsilon/2 < \gamma$. It follows that $(c - \delta, c + \delta) \subset S$, so c cannot be the supremum of S. That is a contradiction. Therefore $f(c) = \gamma$. $\square$

As an application, we prove the following special case of a theorem of Brouwer:

Theorem 5.4.12 *Let $f : [0, 1] \to [0, 1]$ be a continuous function. Then f has a fixed point, in the sense that there is a point $c \in [0, 1]$ such that $f(c) = c$.*

Proof: Seeking a contradiction, we suppose not. Then, in particular, $f(0) > 0$

and $f(1) < 1$. Now set $g(x) = x - f(x)$. We see that $g(0) = 0 - f(0) < 0$ and $g(1) = 1 - f(1) > 0$. By the Intermediate Value Property, there must therefore be a point c between 0 and 1 such that $g(c) = 0$. But this says that $f(c) = c$, as required. □

Recall that, in Section 4.5 we established that the set of all decimal representations of numbers is uncountable. It follows that the set of all real numbers is uncountable. In fact the same proof shows that the set of all real numbers in the interval $(0, 1)$, or in any non-empty open interval (c, d), is uncountable.

The set $\mathbb{R}$ of real numbers is uncountable, yet the set $\mathbb{Q}$ of rational numbers is countable. It follows that the set $\mathbb{R} \setminus \mathbb{Q}$ of *irrational* numbers is uncountable. In particular, it is non-empty. Thus we may see with very little effort that there exist a great many real numbers that cannot be expressed as a quotient of integers. However, given any particular real number (such as π or e or $\sqrt{2}^{\sqrt{2}}$), it can be quite difficult to see whether it is irrational.

We conclude by recalling the "absolute value" notation:

Definition 5.4.13 Let x be a real number. We define

$$|x| = \begin{cases} x & \text{if} \quad x > 0 \\ 0 & \text{if} \quad x = 0 \\ -x & \text{if} \quad x < 0 \end{cases}$$

The absolute value of a real number x measures the distance of x to 0.

The most important property of the absolute value is the *triangle inequality*:

$$|x + y| \leq |x| + |y|.$$

5.5 OPTIONAL SECTION: Construction of the Real Number System

There are several techniques for constructing the real number system $\mathbb{R}$ from the rational number system $\mathbb{Q}$. We use the method of Dedekind (Julius W. R. Dedekind, 1831–1916) cuts because it uses a minimum of new ideas and is fairly brief.

Keep in mind that, throughout this section, our universe is the system of rational numbers $\mathbb{Q}$. We are *constructing* the new number system $\mathbb{R}$.

Definition 5.5.1 A *cut* is a subset $\mathcal{C}$ of $\mathbb{Q}$ with the following properties:

(1) $\mathcal{C} \neq \emptyset$.

(2) If $s \in \mathcal{C}$ and $t < s$, then $t \in \mathcal{C}$.

FIGURE 5.2
A cut.

(3) If $s \in C$, then there is a $u \in C$ such that $u > s$.

(4) There is a rational number x such that $c < x$ for all $c \in C$.

You should think of a cut C as the set of all rational numbers to the left of some point in the real line (that is, it is an open half-line of rational numbers—see Figure 6.2).

For example, the set $\{x \in \mathbb{Q} : x^2 < 2\} \cup \{x \in \mathbb{Q} : x < 0\}$ is a cut. Roughly speaking, it is the set of rational numbers to the left of $\sqrt{2}$. [Take care to note that $\sqrt{2}$ does not exist as a rational number; so we are using a circuitous method to specify this set.] Since we have not constructed the real line yet, we cannot define this cut in that simple way; we have to make the construction more indirect. But if you consider the four properties of a cut, they describe a set that looks like a "rational left half-line."

Notice that if C is a cut and $s \notin C$ then any rational $t > s$ is also not in C. Also, if $r \in C$ and $s \notin C$, then it must be that $r < s$.

Definition 5.5.2 If C and D are cuts, then we say that $C < D$ provided that C is a subset of D but $C \neq D$.

Check for yourself that "<" is a strict, simple ordering on the set of all cuts. We note that $C = D$ if and only if $C \subset D$ and $D \subset C$.

Now we introduce operations of addition and multiplication that will turn the set of all cuts into a field.

Definition 5.5.3 If C and D are cuts, then we define

$$C + D = \{c + d : c \in C, d \in D\}.$$

We define the cut $\hat{0}$ to be the set of all negative rationals.

The cut $\hat{0}$ will play the role of the additive identity. We are now required to check that field axioms A1 - A5 hold.

For **A1**, we need to see that $C + D$ is a cut. Obviously $C + D$ is not empty. If s is an element of $C + D$ and t is a rational number less than s, write $s = c + d$, where $c \in C$ and $d \in D$. Then $t - c < s - c = d \in D$ so $t - c \in D$; and $c \in C$. Hence $t = c + (t - c) \in C + D$. A similar argument shows that there is an $r > s$ such that $r \in C + D$. Finally, if x is a rational upper bound for C and y is a rational upper bound for D, then $x + y$ is a rational upper bound for $C + D$. We conclude that $C + D$ is a cut.

Since addition of rational numbers is commutative, it follows immediately that addition of cuts is commutative. Associativity follows in a similar fashion. That takes care of **A2** and **A3**.

Now we show that if $\mathcal{C}$ is a cut, then $\mathcal{C} + \hat{0} = \mathcal{C}$. For if $c \in \mathcal{C}$ and $z \in \hat{0}$, then $c + z < c + 0 = c$ hence $\mathcal{C} + \hat{0} \subset \mathcal{C}$. Also, if $c^* \in \mathcal{C}$ then choose a $d^* \in \mathcal{C}$ such that $c^* < d^*$. Then $c^* - d^* < 0$, so $c^* - d^* \in \hat{0}$. And $c^* = d^* + (c^* - d^*)$. Hence $\mathcal{C} \subset \mathcal{C} + \hat{0}$. We conclude that $\mathcal{C} + \hat{0} = \mathcal{C}$. This is **A4**.

Finally, for Axiom **A5**, we let $\mathcal{C}$ be a cut and set $-\mathcal{C}$ to be equal to $\{d \in \mathbb{Q} : \exists d^* > d$ such that $c + d^* < 0$ for all $c \in \mathcal{C}\}$. If x is a rational upper bound for $\mathcal{C}$, then $-x \in -\mathcal{C}$ so $-\mathcal{C}$ is not empty. It is also routine to check that $-\mathcal{C}$ is a cut. By its very definition, $\mathcal{C} + (-\mathcal{C}) \subset \hat{0}$.

Further, if $z \in \hat{0}$, then there is a $z^* \in \hat{0}$ such that $z < z^*$. Choose an element $c \in \mathcal{C}$ such that $c + (z^* - z) \notin \mathcal{C}$ (why is this possible?). Let $c^* \in \mathcal{C}$ be such that $c < c^*$. Set $c^{**} = z - c^*$. Then $d^* = z - c > c^{**}$. We claim that $\tilde{c} + d^* < 0$ for all $\tilde{c} \in \mathcal{C}$. Suppose for the moment that this claim has been proved. Then this shows that $c^{**} \in -\mathcal{C}$. Then $z = c^* + c^{**} \in \mathcal{C} + (-\mathcal{C})$ so that $\hat{0} \subset \mathcal{C} + (-\mathcal{C})$. We then conclude that $\mathcal{C} + (-\mathcal{C}) = \hat{0}$, and Axiom **A5** is established.

It remains to prove the claim. So let d^* be defined as above, and select $\tilde{c} \in \mathcal{C}$. Then

$$d^* + \tilde{c} = z + (-c + \tilde{c}) < z + (z^* - z) = z^* < 0.$$

Here we have used the choice of c. This establishes the claim and completes the proof of **A5**.

Having verified the axioms for addition, we turn now to multiplication.

Definition 5.5.4 If $\mathcal{C}$ and $\mathcal{D}$ are cuts, then we define the product $\mathcal{C} \cdot \mathcal{D}$ as follows:

- If $\mathcal{C}, \mathcal{D} > \hat{0}$, then $\mathcal{C} \cdot \mathcal{D} = \{q \in \mathbb{Q} : q < c \cdot d\}$ for some $c \in \mathcal{C}, d \in \mathcal{D}$ with $c > 0, d > 0$.

- If $\mathcal{C} > \hat{0}, \mathcal{D} < \hat{0}$, then $\mathcal{C} \cdot \mathcal{D} = -(\mathcal{C} \cdot (-\mathcal{D}))$.

- If $\mathcal{C} < \hat{0}, \mathcal{D} > \hat{0}$, then $\mathcal{C} \cdot \mathcal{D} = -((-\mathcal{C}) \cdot \mathcal{D})$.

- If $\mathcal{C}, \mathcal{D} < \hat{0}$, then $\mathcal{C} \cdot \mathcal{D} = (-\mathcal{C}) \cdot (-\mathcal{D})$.

- If either $\mathcal{C} = \hat{0}$ or $\mathcal{D} = \hat{0}$, then $\mathcal{C} \cdot \mathcal{D} = \hat{0}$.

Notice that, for convenience, we have defined multiplication of negative numbers just as we did in high school. The reason is that the definition that we use for the product of two positive numbers cannot work when one of the two factors is negative (exercise).

We have said what the additive identity is in this realization of the real numbers. Of course the multiplicative identity is the cut corresponding to 1, or

$$\hat{1} \equiv \{t \in \mathbb{Q} : t < 1\}.$$

We leave it to the reader to verify that if C is any cut, then $\hat{1} \cdot C = C \cdot \hat{1} = C$.

It is now routine to verify that the set of all cuts, with this definition of multiplication, satisfies field axioms **M1 - M5**. The proofs follow those for **A1 - A5** rather closely.

For the distributive property, one first checks the case when all the cuts are positive, reducing it to the distributive property for the rationals. Then one handles negative cuts on a case-by-case basis.

The two properties of an ordered field are also easily checked for the set of all cuts.

We now know that the collection of all cuts forms an ordered field. Denote this field by the symbol $\mathbb{R}$ and call it the *real number system*. We next verify the crucial property of $\mathbb{R}$ that sets it apart from $\mathbb{Q}$:

Theorem 5.5.5 *The ordered field $\mathbb{R}$ satisfies the least upper bound property.*

Proof: Let S be a subset of $\mathbb{R}$ which is bounded above. That is, there is a cut α such that $s < \alpha$ for all $s \in S$. Define

$$ \mathcal{S}^* = \bigcup_{\mathcal{C} \in S} \mathcal{C}. $$

Then $\mathcal{S}^*$ is clearly non-empty, and it is therefore a cut since it is a union of cuts. It is also clearly an upper bound for S since it contains each element of S. It remains to check that $\mathcal{S}^*$ is the least upper bound for S.

In fact, if $\mathcal{T} < \mathcal{S}^*$, then $\mathcal{T} \subset \mathcal{S}^*$ and there is a rational number q in $\mathcal{S}^* \setminus \mathcal{T}$. But, by the definition of $\mathcal{S}^*$, it must be that $q \in \mathcal{C}$ for some $\mathcal{C} \in S$. So $\mathcal{C} > \mathcal{T}$, and $\mathcal{T}$ cannot be an upper bound for S. Therefore $\mathcal{S}^*$ is the least upper bound for S, as desired. $\qquad\square$

We have shown that $\mathbb{R}$ is an ordered field that satisfies the least upper bound property. It remains to show that $\mathbb{R}$ contains (a copy of) $\mathbb{Q}$ in a natural way. In fact, if $q \in \mathbb{Q}$ we associate to it the element $\varphi(q) = \mathcal{C}_q \equiv \{x \in \mathbb{Q} : x < q\}$. Then $\mathcal{C}_q$ is obviously a cut. It is also routine to check that

$$ \varphi(q + q^*) = \varphi(q) + \varphi(q^*) \quad \text{and} \quad \varphi(q \cdot q^*) = \varphi(q) \cdot \varphi(q^*). $$

Therefore we see that φ is a ring homomorphism (see [LAN]) and hence represents $\mathbb{Q}$ as a "subfield" of $\mathbb{R}$.

5.6 The Nonstandard Real Number System

5.6.1 The need for nonstandard numbers

Isaac Newton's calculus was premised on the existence of certain "infinitesimal numbers"—numbers that are positive, smaller than any standard real number, but not zero. Since limits were not understood in Newton's time, infinitesimals served in their stead. But in fact it was just these infinitesimals that called the theory of calculus into doubt. More than a century was expended developing the theory of limits in order to dispel those doubts.

Nonstandard analysis, due to Abraham Robinson (1918–1974), is a model for the real numbers (i.e., it is a number system that satisfies the axioms for the real numbers which we enunciated in Section 5.4) that contains infinitesimals. In a sense, then, Robinson's nonstandard reals are a perfectly rigorous theory that vindicates Newton's original ideas about infinitesimally small numbers.

5.6.2 Filters and ultrafilters

One of the most standard constructions of the nonstandard real numbers involves putting an equivalence relation on the set of all sequences $\{a_j\}$ of real numbers. A natural algebraic construction for doing so is the *ultrafilter*. In fact, ultrafilters are widely used in model theory (see the article by P. C. Eklof in [BAR]). So we shall briefly explain now what an ultrafilter is.

Let I be a non-empty set. A *filter* over I is a set $D \subset \mathcal{P}(I)$ such that

1. $\emptyset \notin D, I \in D$;
2. If $X, Y \in D$, then $X \cap Y \in D$;
3. If $X \in D$ and $X \subset Y \subset I$, then $Y \in D$.

In particular, a filter D over I has the *finite intersection property*: the intersection of any finite set of elements of D is non-empty.

A filter D over I is called an *ultrafilter* if, for every $X \subset I$, either $X \in D$ or $I \setminus X \in D$. It turns out that a filter over I is an ultrafilter if and only if it is a maximal filter over I (i.e., there is no larger filter containing it). One can show, using Zorn's lemma (see the Appendix), that if S is a collection of subsets of I that has the finite intersection property, then S is contained in an ultrafilter over I.

5.6.3 A useful measure

We will follow the exposition that may be found at [NSA]. See also [LIN], [CUT]. At the end, we will point out the ultrafilter that is lurking in the background.

Let m be a finitely additive measure (i.e., a set function) on the set $\mathbb{N}$ of natural numbers such that

1. For any subset $A \subset \mathbb{N}$, $m(A)$ is either 0 or 1.
2. It holds that $m(\mathbb{N}) = 1$ and $m(B) = 0$ for any finite set B.

That such a measure m exists is an easy exercise with the Axiom of Choice (see the Appendix). We leave the details to the interested reader.

5.6.4   An equivalence relation

Let

$$S = \left\{ \{a_n\}_{n=1}^{\infty} : a_n \in \mathbb{R} \text{ for all } n = 1, 2, \ldots \right\}.$$

Define a relation $\sim$ on S by

$$\{a_n\} \sim \{b_n\} \quad \text{if and only if} \quad m\{n : a_n = b_n\} = 1.$$

Then $\sim$ is clearly an equivalence relation. We let $\mathbb{R}^* = S/\sim$ (that is, the collection of equivalence classes) be the nonstandard real number system.[1]

We let $[\{a_n\}]$ denote the equivalence class containing the sequence $\{a_n\}$. Then we define some of the elementary operations on $\mathbb{R}^*$ by

$$[\{a_n\}] + [\{b_n\}] = [\{a_n + b_n\}];$$

$$[\{a_n\}] \cdot [\{b_n\}] = [\{a_n \cdot b_n\}];$$

$$[\{a_n\}] < [\{b_n\}] \quad \text{iff} \quad m(\{n : a_n < b_n\}) = 1.$$

Further, we identify a standard real number b with the equivalence class $[\{b, b, b, \ldots\}]$.

5.6.5   An extension of the real number system

We have seen that $\mathbb{R}^*$ clearly contains $\mathbb{R}$ in a natural way. And it contains other elements too. We call $x \in \mathbb{R}^*$ an *infinitesimal* if and only if $a \neq 0$ and $-a < x < a$ for every positive real number a. For example, $[\{1, 2/3, 1/3, \ldots\}]$ is an infinitesimal. We call $y \in \mathbb{R}^*$ an *infinitary number* if $y > b$ for every real number b or $y < d$ for every real number d. As an instance, $[\{1, 2, 3, \ldots\}]$ is an infinitary number.

It would be inappropriate in a book of this type to delve very far into the theory of the nonstandard reals. But at least now the reader has an idea of what the nonstandard real numbers are, and of how a number system could contain both the standard reals and also infinitesimals and infinitaries.

We close this section by noting that John Horton Conway (1937–   ) has trumped Abraham Robinson by creating a new number system called the *surreal numbers*. The surreal numbers form the largest possible ordered field. They contain many other known numbers systems, including the transfinite ordinal numbers. It was Donald E. Knuth (1938–   ) who actually named and developed this number system in [KNU]. You can read about the surreal numbers in Chapter 11 of [KRA2]. See also

https://mathworld.wolfram.com/NonstandardAnalysis.html .

---

[1]In fact this is the point where we use an ultrafilter. The set $\mathcal{M} = \{A \subset \mathbb{N} : m(A) = 1\}$ is an ultrafilter. We are mod-ing out by this ultrafilter.

Abraham Robinson
1918–1974

Abraham Robinson was educated in Palestine. He began studying mathematics under Fraenkel and Levitzki at the Hebrew University of Jerusalem. Robinson was a brilliant student and, after graduating in 1939, he was awarded a scholarship to allow him to study at the Sorbonne in Paris.

As an undergraduate at the Hebrew University, Robinson was interested in both algebra and mathematical logic. However, once in England he enlisted in the Free French Air Force and, because he was a mathematician, he was sent in 1941 to the Royal Aircraft Establishment at Farnborough where he became a Scientific Officer.

He went to the University of Toronto in 1951 to take up a chair of applied mathematics but left for Jerusalem in 1957 to fill Fraenkel's chair at the Hebrew University. He was Chairman of the Mathematics Department there until 1962 when he accepted the professorship of Mathematics and Philosophy at the University of California, Los Angeles. In 1967 he moved again, but remaining in the United States he went to Yale University as Professor of Mathematics. The sudden fatal illness of Abraham Robinson came as a great shock to many people around the world. For Robinson was more than an excellent mathematician. He was also a person whom one came very quickly to like very much.

Robinson is best known as the inventor of the nonstandard real numbers. But he also made substantial contributions to algebra, analysis, and applied mathematics. He was a scholar of many and diverse interests.

5.7 The Complex Numbers

When we first learn about the complex numbers, the most troublesome point is the very beginning: "Let's pretend that the number -1 has a square root. Call it i." What gives us the right to "pretend" in this fashion? The answer is that we have no such right. If -1 has a square root, we should be able to construct a number system in which that is the case. That is what we shall do in this section.

Definition 5.7.1 The system of *complex numbers*, denoted by the symbol $\mathbb{C}$, consists of all ordered pairs (a, b) of real numbers (in other words, $\mathbb{C} = \mathbb{R} \times \mathbb{R}$). We add two complex numbers (a, b) and (a^*, b^*) by the formula

$$(a, b) + (a^*, b^*) = (a + a^*, b + b^*).$$

We multiply two complex numbers by the formula

$$(a, b) \cdot (a^*, b^*) = (a \cdot a^* - b \cdot b^*, a \cdot b^* + a^* \cdot b).$$

Remark 5.7.2 If you are puzzled by this definition of multiplication, then do not worry. In a few moments you will see that it gives rise to the notion of multiplication of complex numbers that you have seen before. In any event, this notion of multiplication has the properties we need, and gives the answers that we want.

It is interesting to note that, unlike the integers and the rational numbers, the new number system $\mathbb{C}$ is *not* a collection of equivalence classes. Instead, $\mathbb{C}$ is the Euclidean plane equipped with some new algebraic operations.

EXAMPLE 5.7.3 Let $z = (3, -2)$ and $w = (4, 7)$ be two complex numbers. Then

$$z + w = (3, -2) + (4, 7) = (3 + 4, -2 + 7) = (7, 5).$$

Also

$$z \cdot w = (3, -2) \cdot (4, 7) = (3 \cdot 4 - (-2) \cdot 7, \, 3 \cdot 7 + (-2) \cdot 4) = (26, 13).$$

◇

Donald Knuth (1938–   )

When Donald Knuth was young he entered a competition, with the candy manufacturer Ziegler, to see how many words could be made with the letters of "Ziegler's Giant Bar." Using a dictionary, he came up with 4500 words. Knuth was an easy winner.

Knuth enrolled at Case Western Reserve University in 1956 to study physics. From his second year on, Knuth started to move toward mathematics and away from physics.

In fact Knuth already had his first encounter with computers in his first year at Case. He consulted the IBM 650 manual to find out how to write programs: "the manual we got from IBM would show examples of programs and I knew I could do better than that. So I thought I might have some talent."

In the autumn of 1960, Knuth entered the California Institute of Technology and, in June 1963, he was awarded a Ph.D. in mathematics. In fact, in addition to the work for his doctorate in mathematics, Knuth had from 1960 became a software consultant to the Burroughs Corporation in Pasadena, California.

After completion of his doctorate in 1963, Knuth became an Assistant Professor of Mathematics at the California Institute of Technology, being promoted to Associate Professor in 1966.

By 1966 his book, long in development, on compilers had grown to 3000 handwritten pages. Discussions with the publisher led to a decision that Knuth should produce a seven-volume work covering much more than compilers. The work became *The Art of Computer Programming* and publication began in 1968 when Volume 1: Fundamental Algorithms appeared. Volume 2: Seminumerical Algorithms came out in the following year, and Volume 3: Sorting and Searching in 1973.

In 1968, Knuth was appointed as Professor of Computer Science at Stanford University. Knuth remained at Stanford University for the remainder of his career.

Knuth has made many contributions to mathematics and computing. One of the most remarkable is Knuth's invention of TeX, a language for typesetting mathematical and scientific articles. Starting in 1976, Knuth took ten years off his other projects to work on the development of TeX and METAFONT, a computer software system for font design.

TeX has changed mathematics publishing since it enables mathematicians to produce high-quality mathematical documents using a home computer. However, it has not only changed the way that mathematical and scientific articles are published but also in the way that they are communicated.

We should mention a few of the many contributions by Knuth: semantics of programming languages; attribution grammar; the development of $LR(k)$ parsing; the Knuth-Morris-Pratt algorithm which searches for a string of characters; and structured documentation and literate programming.

For his quite remarkable contributions Knuth has received many honours, too numerous to list here.

As usual, we ought to check that addition and multiplication are commutative, associative, that multiplication distributes over addition, and so forth. We shall leave these tasks to the exercises. Instead we develop some of the crucial properties of our new number system.

Theorem 5.7.4 *The following properties hold for the number system* $\mathbb{C}$.

> **(1)** *The number* $1 \equiv (1,0)$ *is the multiplicative identity:* $1 \cdot z = z$ *for any* $z \in \mathbb{C}$.
>
> **(2)** *The number* $0 \equiv (0,0)$ *is the additive identity:* $0 + z = z$ *for any* $z \in \mathbb{C}$.
>
> **(3)** *Each complex number* $z = (x,y)$ *has an additive inverse* $-z = (-x,-y)$: *it holds that* $z + (-z) = 0$.
>
> **(4)** *The number* $i \equiv (0,1)$ *satisfies* $i \cdot i = (-1,0) \equiv -1$; *in other words,* i *is a square root of* -1.

Proof: These are direct calculations, but it is important for us to work out these facts.

First, let $z = (x,y)$ be any complex number. Then

$$1 \cdot z = (1,0) \cdot (x,y) = (1 \cdot x - 0 \cdot y, 1 \cdot y + 0 \cdot x) = (x,y) = z.$$

This proves the first assertion.

For the second assertion, we have

$$0 + z = (0,0) + (x,y) = (0 + x, 0 + y) = (x,y) = z.$$

With z as above, set $-z = (-x,-y)$. Then

$$z + (-z) = (x,y) + (-x,-y) = (x + (-x), y + (-y)) = (0,0) = 0.$$

Finally, we calculate

$$i \cdot i = (0,1) \cdot (0,1) = (0 \cdot 0 - 1 \cdot 1, 0 \cdot 1 + 1 \cdot 0) = (-1,0) = -1.$$

Thus, as asserted, i is a square root of -1. $\square$

Proposition 5.7.5 *If* $z \in \mathbb{C}, z \neq 0$, *then there is a complex number* w *such that* $z \cdot w = 1$.

Proof: Write $z = (x,y)$ and set

$$w = \left(\frac{x}{x^2 + y^2}, \frac{-y}{x^2 + y^2} \right).$$

Since $z \neq 0$, this definition makes sense. Then it is straightforward to verify that $z \cdot w = 1$. Indeed,

$$
\begin{aligned}
z \cdot w &= (x, y) \cdot \left(\frac{x}{x^2 + y^2}, \frac{-y}{x^2 + y^2} \right) \\
&= \left(\frac{x \cdot x}{x^2 + y^2} - \frac{y \cdot (-y)}{x^2 + y^2}, \frac{x \cdot (-y)}{x^2 + y^2} + \frac{y \cdot x}{x^2 + y^2} \right) \\
&= (1, 0) \, . \qquad\qquad\qquad\qquad\qquad\qquad\qquad \square
\end{aligned}
$$

Thus every non-zero complex number has a multiplicative inverse. The other field axioms for $\mathbb{C}$ are easy to check. We conclude that the number system $\mathbb{C}$ forms a field. You will prove in the exercises that it is not possible to order this field. If α is a real number, then we associate α with the complex number $(\alpha, 0)$. In this way, we can think of the real numbers as a *subset* of the complex numbers. In fact, the real field $\mathbb{R}$ is a *subfield* of the complex field $\mathbb{C}$. This means that if $\alpha, \beta \in \mathbb{R}$ and $(\alpha, 0), (\beta, 0)$ are the corresponding elements in $\mathbb{C}$, then $\alpha + \beta$ corresponds to $(\alpha + \beta, 0)$ and $\alpha \cdot \beta$ corresponds to $(\alpha, 0) \cdot (\beta, 0) = (\alpha\beta, 0)$.

These assertions are explored more thoroughly in the exercises.

With the remarks in the preceding paragraphs, we can sometimes ignore the distinction between the real numbers and the complex numbers. For example, we can write

$$ 5 \cdot i $$

and understand that it means $(5, 0) \cdot (0, 1) = (0, 5)$. Likewise, the expression

$$ 5 \cdot 1 $$

can be interpreted as $5 \cdot 1 = 5$ or as $(5, 0) \cdot (1, 0) = (5, 0)$ without any danger of ambiguity or misunderstanding.

Theorem 5.7.6 *Every complex number can be written in the form $a + b \cdot i$, where a and b are real numbers. In fact, if $z = (a, b) \in \mathbb{C}$, then*

$$ z = a + b \cdot i. $$

Proof: With the identification of real numbers as a subfield of the complex numbers, we have that

$$ a + b \cdot i = (a, 0) + (b, 0) \cdot (0, 1) = (a, 0) + (0, b) = (a, b) = z $$

as claimed. $\qquad\qquad\qquad\qquad\qquad\qquad\qquad\qquad\qquad\qquad\qquad\qquad \square$

Now that we have constructed the complex number field, we will adhere to the usual custom of writing complex numbers as $z = a + b \cdot i$ or, more simply,

$a + bi$. We call a the *real part* of z, denoted by Re z, and b the *imaginary part* of z, denoted Im z. In this notation, our algebraic operations become

$$(a + bi) + (a^* + b^*i) = (a + a^*) + (b + b^*)i$$

and

$$(a + bi) \cdot (a^* + b^*i) = (a \cdot a^* - b \cdot b^*) + (a \cdot b^* + a^* \cdot b)i.$$

If $z = a + bi$ is a complex number, then we define its *complex conjugate* to be the number $\bar{z} = a - bi$. We record some elementary facts about the complex conjugate:

Proposition 5.7.7 *If z, w are complex numbers, then*

  (1) $\overline{z + w} = \bar{z} + \bar{w}$;

  (2) $\overline{z \cdot w} = \bar{z} \cdot \bar{w}$;

  (3) $z + \bar{z} = 2 \cdot \operatorname{Re} z$;

  (4) $z - \bar{z} = 2 \cdot i \cdot \operatorname{Im} z$;

  (5) $z \cdot \bar{z} \geq 0$, *with equality holding if and only if $z = 0$.*

Proof: Write $z = a + bi, w = c + di$. Then

$$
\begin{aligned}
\overline{z + w} &= \overline{(a + c) + (b + d)i} \\
&= (a + c) - (b + d)i \\
&= (a - bi) + (c - di) \\
&= \bar{z} + \bar{w}.
\end{aligned}
$$

This proves **(1)**. Assertions **(2)**, **(3)**, **(4)** are proved similarly. For **(5)**, notice that

$$z \cdot \bar{z} = (a + bi) \cdot (a - bi) = a^2 + b^2 \geq 0.$$

Clearly equality holds if and only if $a = b = 0$.   □

The expression $|z|$ is defined to be the nonnegative square root of $z \cdot \bar{z}$. In other words

$$|z| = \sqrt{z \cdot \bar{z}} = \sqrt{(x + iy) \cdot (x - iy)} = \sqrt{x^2 + y^2}.$$

It is called the *modulus* of z and plays the same role for the complex field that absolute value plays for the real field: the modulus of z measures the distance of z to the origin.

The modulus has the following properties.

Proposition 5.7.8 *If $z, w \in \mathbb{C}$, then*

  (1) $|z| = |\bar{z}|$;

(2) $|z \cdot w| = |z| \cdot |w|$;

(3) $|\text{Re } z| \le |z|$, $|\text{Im } z| \le |z|$;

(4) $|z + w| \le |z| + |w|$.

Proof: Write $z = a + bi, w = c + di$. Then **(1)**, **(2)**, **(3)** are immediate. For **(4)** we calculate that

$$
\begin{aligned}
|z + w|^2 &= (z + w) \cdot (\overline{z + w}) \\
&= z \cdot \overline{z} + z \cdot \overline{w} + w \cdot \overline{z} + w \cdot \overline{w} \\
&= |z|^2 + 2\text{Re } (z \cdot \overline{w}) + |w|^2 \\
&\le |z|^2 + 2|z \cdot \overline{w}| + |w|^2 \\
&= |z|^2 + 2|z| \cdot |w| + |w|^2 \\
&= (|z| + |w|)^2.
\end{aligned}
$$

Taking square roots proves **(4)**. $\qquad\qquad\square$

Observe that the fourth part of the Proposition reduces, in the real case, to the triangle inequality

$$|x + y| \le |x| + |y|.$$

The most important property of the complex numbers $\mathbb{C}$ is that $\mathbb{C}$ is *algebraically complete*. This means that any polynomial

$$p(z) = a_0 + a_1 z + a_2 z^2 + \cdots + a_k z^k$$

with complex coefficients has a complex root.

This result was first proved by Carl Friedrich Gauss (1777–1855) in his doctoral dissertation.

Carl Friedrich Gauss (1777–1855)

At the age of seven, Carl Friedrich Gauss started elementary school, and his potential was noticed almost immediately. After receiving a stipend from the Duke of Brunswick-Wolfenbüttel, Gauss entered Brunswick Collegium Carolinum in 1792.

In 1795, Gauss left Brunswick to study at Göttingen University. Gauss left Göttingen in 1798 without a diploma, but by this time he had made one of his most important discoveries—the construction of a regular 17-gon by ruler and compasses.

Gauss returned to Brunswick where he received a degree in 1799. After the Duke of Brunswick had agreed to continue Gauss's stipend, he requested that Gauss submit a doctoral dissertation to the University of Helmstedt. He already knew Pfaff, who was chosen to be his advisor. Gauss's dissertation was a discussion of the fundamental theorem of algebra.

With his stipend to support him, Gauss did not need to find a job so devoted himself to research. He published the book *Disquisitiones Arithmeticae* in the summer of 1801.

Much of Gauss's time was spent on a new observatory, completed in 1816, but he still found the time to work on other subjects. Gauss found himself more and more interested in geodesy in the 1820s. Gauss had been asked in 1818 to carry out a geodesic survey of the state of Hanover to link up with the existing Danish grid. Because of the survey, Gauss invented the heliotrope which worked by reflecting the Sun's rays using a design of mirrors and a small telescope.

In 1822, Gauss won the Copenhagen University Prize. From the early 1800s, Gauss had an interest in the question of the possible existence of a non-Euclidean geometry.

In 1832, Gauss and Weber began investigating the theory of terrestrial magnetism after Alexander von Humboldt attempted to obtain Gauss's assistance in making a grid of magnetic observation points around the Earth.

In 1837, Weber was forced to leave Göttingen when he became involved in a political dispute and, from this time, Gauss's activity decreased. He still produced letters in response to fellow scientists' discoveries usually remarking that he had known the methods for years but had never felt the need to publish.

Gauss presented his golden jubilee lecture in 1849, fifty years after his diploma had been granted by Helmstedt University. From the mathematical community only Jacobi and Dirichlet were present, but Gauss received many messages and honours. His health deteriorated slowly, and Gauss died in his sleep early in the morning of 23 February, 1855.

EXAMPLE 5.7.9 Let us find all the roots of the polynomial

$$p(z) = z^3 + (-3 - i)z^2 + (2 + 3i)z - 2i.$$

With some experimentation, we find that $z = 1$ is a root of p. Dividing p by $(z - 1)$, we find that

$$P(z) = (z - 1) \cdot (z^2 + (-2 - i)z + 2i).$$

Of course we can solve the quadratic polynomial using the quadratic formula. We find the two additional roots 2 and i. ◇

Of course this result, the Fundamental Theorem of Algebra, is a deep and difficult theorem. We shall treat its proof in Exercise 6.47.

5.8 The Quaternions, the Cayley Numbers, and Beyond

Now we shall discuss a number system that you may have never encountered before. It is called the system of *quaternions*. Our description will be an informal one.

The quaternions were discovered by William Rowan Hamilton (1805–1865). He had been trying for many years to produce a field structure on $\mathbb{R}^3$. One day, walking to an important meeting with his wife, he realized that he should not be wedded to commutativity of multiplication, and also not be wedded to dimension 3. He was so excited by his discovery that he whipped out his penknife and carved the result onto the bridge that they were crossing. So now we have the quaternions. You might think it natural to denote the quaternions by $\mathbb{Q}$. But that notation is already reserved for the rational numbers. Instead we denote the quaternions by $\mathbb{H}$, in honor of Hamilton.

Imagine $\mathbb{R}^4 \equiv \mathbb{R} \times \mathbb{R} \times \mathbb{R} \times \mathbb{R}$ equipped with the following operations: set $\mathbf{i} = (0, 1, 0, 0)$, $\mathbf{j} = (0, 0, 1, 0)$, $\mathbf{k} = (0, 0, 0, 1)$. Denote the 4-tuple $(1, 0, 0, 0)$ by $\mathbf{1}$. Define the multiplication laws

$$\mathbf{i} \cdot \mathbf{i} = -1 \quad , \quad \mathbf{j} \cdot \mathbf{j} = -1 \quad , \quad \mathbf{k} \cdot \mathbf{k} = -1$$

and

$$\mathbf{i} \cdot \mathbf{j} = \mathbf{k} \quad , \quad \mathbf{j} \cdot \mathbf{k} = \mathbf{i} \quad , \quad \mathbf{k} \cdot \mathbf{i} = \mathbf{j}$$

and

$$\mathbf{j} \cdot \mathbf{i} = -\mathbf{k} \quad , \quad \mathbf{k} \cdot \mathbf{j} = -\mathbf{i} \quad , \quad \mathbf{i} \cdot \mathbf{k} = -\mathbf{j}.$$

Of course the element $\mathbf{1}$ multiplied times any 4-tuple z is declared to be equal to z. In particular, $\mathbf{1} \cdot \mathbf{1} = \mathbf{1}$.

Finally, if $z = (z_1, z_2, z_3, z_4)$ and $w = (w_1, w_2, w_3, w_4)$ are 4-tuples, then we write

$$z = z_1 \cdot \mathbf{1} + z_2 \mathbf{i} + z_3 \mathbf{j} + z_4 \mathbf{k}$$

and

$$w = w_1 \cdot \mathbf{1} + w_2 \mathbf{i} + w_3 \mathbf{j} + w_4 \mathbf{k}.$$

Then $z \cdot w$ is defined by using the (obvious) distributive law and the rules already specified. For example,

$$
\begin{aligned}
(2,0,1,3) \cdot (-4,1,0,1) &= [2 \cdot \mathbf{1} + \mathbf{j} + 3\mathbf{k}] \cdot [-4 \cdot \mathbf{1} + \mathbf{i} + \mathbf{k}] \\
&= (2 \cdot (-4)) \cdot \mathbf{1} + (2\mathbf{i}) + (2\mathbf{k}) \\
&\quad + (\mathbf{j} \cdot (-4)) + (\mathbf{j} \cdot \mathbf{i}) + (\mathbf{j} \cdot \mathbf{k}) \\
&\quad + (3\mathbf{k} \cdot (-4)) + (3\mathbf{k} \cdot \mathbf{i}) + (3\mathbf{k} \cdot \mathbf{k}) \\
&= -8 \cdot \mathbf{1} + 2\mathbf{i} + 2\mathbf{k} - 4\mathbf{j} - \mathbf{k} + \mathbf{i} \\
&\quad - 12\mathbf{k} + 3\mathbf{j} - 3 \cdot \mathbf{1} \\
&= -11 \cdot \mathbf{1} + 3\mathbf{i} - \mathbf{j} - 11\mathbf{k} \\
&= (-11, 3, -1, -11).
\end{aligned}
$$

Addition of two quaternions is simply performed componentwise: if $z = (z_1, z_2, z_3, z_4)$ and $w = (w_1, w_2, w_3, w_4)$, then

$$
z + w = (z_1 + w_1, z_2 + w_2, z_3 + w_3, z_4 + w_4).
$$

Verify for yourself that the additive identity in the quaternions is $(0,0,0,0)$. The multiplicative identity is $\mathbf{1} = (1,0,0,0)$.

In fact it can be checked that each non-zero element of the quaternions has a unique two-sided multiplicative inverse. However, since multiplication is not commutative, the quaternions do not form a field; instead the algebraic structure is called a *division ring*.

It is also possible to give $\mathbb{R}^8$ an additive and a multiplicative structure. The multiplication operation is both noncommutative and non-associative. The resulting eight dimensional algebraic object is called the *Cayley numbers* or *octonions*. We shall not present the details here. It is one of the great theorems of twentieth-century mathematics (see [ADA], [BOM]) that $\mathbb{R}^1$, $\mathbb{R}^2$, $\mathbb{R}^4$, and $\mathbb{R}^8$ are the only Euclidean spaces that can be equipped with compatible addition and multiplication operations in a natural way (so that the algebraic operations are smooth functions of the coordinates).

The quaternions and Cayley numbers are used in mathematical physics, in the representation theory of groups, and in algebraic topology. Every cell phone uses the Cayley numbers for coding.

See https://mathworld.wolfram.com/Octonion.html for more about the Cayley numbers.

Exercises

    1. Let S be a set and let $p : S \times S \to S$ be a binary operation. If $T \subset S$, then we say that T is *closed* under p if $p : T \times T \to T$. [As an example, let $S = \mathbb{Z}$ and T be the even integers and p be ordinary addition.]

Now let $S = \mathbb{R}$. Under which arithmetic operations $+, -, \cdot, \div$ is the set $T = \mathbb{Q}$ closed? Under which arithmetic operations $+, -, \cdot, \div$ is the set $T = \mathbb{R} \setminus \mathbb{Q}$ closed?

2. Imitate the construction of equivalence classes of $\mathbb{Z} \times (\mathbb{Z} \setminus \{0\})$, using $\mathbb{Q} \times (\mathbb{Q} \setminus \{0\})$ instead, as in Section 6.3. What sort of number system results? [**Hint:** Pick a useful representative from each equivalence class.]

3. Imitate the construction of equivalence classes of $\mathbb{N} \times \mathbb{N}$, using $\mathbb{Z} \times \mathbb{Z}$ instead, as in Section 6.2. What sort of number system results? [**Hint:** Pick a useful representative from each equivalence class.]

4. Let q be a rational number. Construct a sequence $\{x_j\}$ of irrational numbers such that $x_j \to q$. This means that, for each $\varepsilon > 0$, there is a positive integer K such that if $j > K$, then $|x_j - q| < \varepsilon$.

5. The numbers $\pi \approx 3.14159\ldots$ and $e \approx 2.71828\ldots$ are both known to be irrational (these assertions are quite difficult to prove). It is unknown whether $x = \pi + e$ or $y = \pi - e$ is irrational. However at least one of these two numbers *must* be irrational. Can you explain why?

6. Let S be a set of real numbers with the property that, whenever $x, y \in S$ and $x < t < y$, then $t \in S$. Can you give a simple description of the set S?

7. Let $a_1 < a_2 < \cdots$ be real numbers. Prove that either there is a real number α such that $a_j \to \alpha$ (refer to Exercise 6.4 for this notation) *or else* the sequence $\{a_j\}$ increases without bound.

8. Let $\{a_j\}_{j=1}^{\infty}$ be a set of real numbers. Let $M > 0$ and assume that $|a_j| \leq M$ for every j. Prove that there is a subsequence $\{a_{j_k}\}$ and a real number x such that for every $\varepsilon > 0$ there is a $K > 0$ such that $|a_{j_k} - x| < \varepsilon$ whenever $k > K$. Refer to Exercises 6.4 and 6.7 for related ideas.

9. Prove that subtraction is well defined in the integers.

10. Prove that multiplication is well defined in the integers.

11. Give a careful discussion of the failure of the operation of division in the integers.

* 12. If x is a positive real number, then let (x) denote its fractional part. For instance, $(3.2) = .2$, $(4/3) = 1/3$, etc.

Now fix an irrational number λ between 0 and 1. Let $a_k = (k\lambda)$. Then each a_k an element of $I = \{t : 0 \leq t < 1\}$. Prove the following: if $\varepsilon > 0$ and $x \in I$ then there is an element a_k such that $|a_k - x| < \varepsilon$. This is a famous result of Herman Weyl.

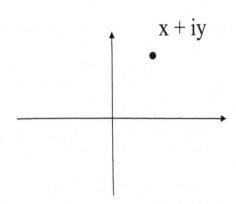

FIGURE 5.3
An Argand diagram.

* **13.** Consider all sequences $\{a_1, a_2, \dots\}$ of rational numbers that satisfy the following condition: If $\varepsilon > 0$, then there exists an integer $N > 0$ such that if $j, k > N$ then $|a_j - a_k| < \varepsilon$. Such a sequence is called a *Cauchy sequence.*

  Say that two Cauchy sequences $\{a_j\}$ and $\{a_j^*\}$ are related if, for any $\varepsilon > 0$, there is a positive integer N such that $j, k > N$ implies $|a_j - a_k^*| < \varepsilon$. Show that this is an equivalence relation. Explain why the set of all equivalence classes is, in a natural way, a model for the real numbers.

14. Prove that addition and subtraction are well defined in the rational number system $\mathbb{Q}$.

15. Determine whether $\sqrt{2} + \sqrt{3}$ is rational or irrational.

* **16.** It is not necessary to discuss well definedness of addition and multiplication in the real number system. Explain why this is so.

17. It is not necessary to discuss well definedness of addition and multiplication in the complex number system. Explain why this is so.

18. Prove that every non-zero complex number $z \in \mathbb{C}$ has two distinct square roots in $\mathbb{C}$.

19. Prove that addition of integers is associative.

20. Assuming that it is known that addition of integers is both commutative and associative, prove then that addition of rational numbers is commutative and associative.

21. An Argand diagram is a device for sketching a complex number in the plane. If $x + iy$ is a complex number then we depict it in the cartesian plane as the point (x, y). See Figure 6.3. Sketch the complex numbers $3 - 2i$, $7 + 4i$, $e + \pi i$, $-6 - i$.

22. Consider the function $f(x) = 2^x$. What does it mean? When x is an integer the meaning is obvious. What about when x is a rational number? What can you say about $f(x)$ when x is an irrational number?

23. The complex number $1 = 1 + 0i$ has three cube roots. Use any means to find them, and sketch them on an Argand diagram (refer to Exercise 6.21 for terminology).

* **24.** Let θ be any real number. A famous formula of Euler asserts that

$$e^{i\theta} = \cos\theta + i\sin\theta.$$

A rigorous verification of this formula requires a study of complex power series (see, for instance, [KRA1]). This exercise provides you with an intuitive argument that should make you comfortable with Euler's formula.

If z is *any* complex number, then define

$$e^z = \sum_{j=0}^{\infty} \frac{z^j}{j!}.$$

Notice that, when z happens to be a real number, then the formula is one that you learned in calculus. The new formula is a standard generalization of the calculus formula. Substitute in $i\theta$ for z and (manipulating the series just as though it were a polynomial) separate the right-hand side into its real and imaginary parts. The result is

$$e^{i\theta} = \left(1 - \frac{\theta^2}{2!} + \frac{\theta^4}{4!} - + \cdots\right) + i\left(\theta - \frac{\theta^3}{3!} + \frac{\theta^5}{5!} - + \cdots\right).$$

Finally, notice that the power series expansions in the parentheses on the right are those associated with the functions cosine and sine, respectively. This is Euler's formula.

25. Refer to Exercise 5.24. If $\xi = s + it$ is any complex number such that $s^2 + t^2 = 1$, then we may find an angle θ, $0 \le \theta < 2\pi$, such that $\cos\theta = s$ and $\sin\theta = t$. See Figure 5.4. We conclude that

$$\xi = e^{i\theta}.$$

Explain this reasoning in detail.

26. If $z = x + iy \in \mathbb{C}$ is any nonzero complex number, then let

$$r^2 = |z|^2 = x^2 + y^2.$$

The number r is the distance of z to the origin in the Argand plane

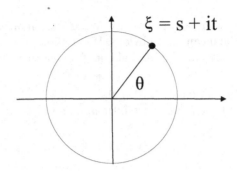

FIGURE 5.4
The angle associated to a complex number of modulus 1.

(Exercises 5.21). It is also the modulus of z. Set $\xi = z/r$. Show that $|\xi| = 1$. Now apply Exercise 5.25 to conclude that

$$z = r\,e^{i\theta},$$

some $0 \leq \theta < 2\pi$. This is called the *polar form* of the complex number z.

27. If $z = re^{i\theta}$ is a complex number in polar form, then $re^{i(\theta+2\pi)}$ is the same complex number (refer to Exercises 5.24, 5.25, 5.26). Explain.

28. Let z be any non-zero complex number, and let k be an integer exceeding 1. In this exercise we learn to find all the kth roots of z. There will be k of them.

    First we solve the equation

    $$w^k = z = re^{i\theta},$$

    where we have written z in the polar form $z = re^{i\theta}$ (refer to Exercises 5.24, 5.25, 5.26, 5.27). If $w = se^{i\psi}$, then we see that $s = r^{1/k}$ and that $\psi = \theta/k$.

    But we may also solve the equation

    $$w^k = z = r^{i(\theta+2\pi)}$$

    to find a second k^{th} root with $s = r^{1/k}$ and $\psi = \theta/k + 2\pi/k$ (refer to Exercise 5.27).

    Continue this procedure to find $(k-2)$ more k^{th} roots of z.

    Apply the procedure developed here to find all sixth roots of 2. Sketch all six roots, together with the number $2 = 2 + 0i$, on an Argand diagram.

29. Apply the ideas developed in Exercise 5.28 to find all fourth roots of the complex number $1 + i$. Sketch all the roots, together with the number $1 + i$, on an Argand diagram.

30. Apply the ideas developed in Exercise 5.28 to find all sixth roots of the complex number $2 - 3i$. Sketch all the roots, together with the number $2 - 3i$, on an Argand diagram.

31. Apply the ideas developed in Exercise 5.28 to find all third roots of the complex number $-i$. Sketch all the roots, together with the number $-i$, on an Argand diagram.

32. The mapping

$$x + iy \mapsto x + yi + 0j + 0k$$

shows that the complex numbers form a natural algebraic sub-object inside the quaternions. Explain this statement.

33. Prove that the complex numbers cannot be made into an ordered field (as discussed in Section 5.7).

* 34. Discuss the relationship between well ordering and the Axiom of Regularity (begin by considering the natural numbers).

35. Prove those parts of Theorem 5.3.6 that were not proved in the text.

36. Prove those parts of Theorem 5.3.9 that were not proved in the text.

37. Complete the calculation in the proof of Theorem 5.4.8.

38. Prove the generalization of Theorem 5.4.8 that is stated after the proof of that theorem.

39. Prove that addition and multiplication are commutative in the complex number system.

40. Prove that multiplication distributes over addition in the complex number system.

41. Exhibit two distinct ways in which the complex numbers may be viewed as a subfield of the quaternions.

42. In how many different ways can the real numbers be exhibited as a subfield of the quaternions?

43. Let z be an element of the quaternions. If $z = z_1 \cdot 1 + z_2 i + z_3 j + z_4 k$, then define $\bar{z} = z_1 \cdot 1 - z_2 i - z_3 j - z_4 k$. Calculate $z \cdot \bar{z}$. In what sense is this bar operation analogous to conjugation in the complex number system?

44. Let $\mathcal{M}_{n \times n}$ denote the set of $n \times n$ matrices with real entries, equipped with the usual matrix addition and multiplication. Explain why $\mathcal{M}_{n \times n}$ does not form a field when equipped with standard matrix addition and multiplication.

45. Let p be a polynomial and assume that $\alpha \in \mathbb{C}$ is a root of p. Prove that $(z - \alpha)$ evenly divides $p(z)$ with no remainder.

46. One of the most crucial properties of the complex number system is enunciated by the Fundamental Theorem of Algebra. This theorem states that if $p(z)$ is a polynomial of degree at least one, then p has a root in the complex number field. That is, there is a number $\alpha \in \mathbb{C}$ such that $p(\alpha) = 0$.

   Assume the Fundamental Theorem of Algebra for the moment. Use the result of Exercise 5.45 to show that if $p(z)$ is a polynomial of degree $k \geq 1$, then p has k roots (possibly not all distinct).

* **47.** Here is a sketch of a proof of the Fundamental Theorem of Algebra (Exercise 5.46). Fill in as many details of the proof as you can. Do not be afraid to consult your instructor for help.

   (a) Let $p(z) = a_0 + a_1 z + a_2 z^2 + \cdots + a_k z^k$ be the polynomial under consideration. We are assuming that $k \geq 1$.

   (b) If $a_0 = 0$ then 0 is a root of the polynomial (why?). So we may as well only consider the case in which $a_0 \neq 0$.

   (c) Fix a positive number R and let C_R denote the circle in the complex plane with center at the origin and radius R. Let us consider the image of C_R under the polynomial p (think of p as a mapping from $\mathbb{C}$ to $\mathbb{C}$).

   (d) If R is very large then we can study the situation in part (c) by writing the polynomial as

   $$p(z) = z^k \cdot \left(a_k + a_{k-1}\frac{1}{z} + \cdots + a_1\frac{1}{z^{k-1}} + a_0\frac{1}{z^k} \right).$$

   When $z \in C_R$ and $R > 0$ is very large, then each of the terms in parentheses, except for the first, is very small. Thus, for such z, the expression $p(z)$ is approximately equal to $z^k \cdot a_k$. The image of C_R under this last expression is a large circle that surrounds the origin.

   (e) If R is very small, then (returning to the original expression for $p(z)$) the quantity $p(z)$ approximately equals $a_0 \neq 0$—for the other terms in the polynomial will be small too. Thus the image of C_R under the mapping p will be some small curve near the point $a_0 \neq 0$. In particular, if R is small enough, then this small curve *will not* surround the origin.

   (f) We see from parts (d) and (e) that when R is small, then the image of C_R under p is a curve that *does not* surround the origin; but if R is large, then the image of C_R under p is a curve that *does* surround the origin.

   (g) It follows from continuity considerations that there must be an intermediate value of R (refer to part (f)) such that the image of C_R under p is a curve that passes through the origin. In other words, for that special value of R, there is a point z_0 on the curve C_R such that $p(z_0) = 0$. But then z_0 is the root that we seek.

48. The Division Algorithm states that, if p, q are positive integers with $p < q$, then we may write $q = p \cdot k + s$, where k is a positive integer and $s < p$. This is just the familiar algorithm of division that we learn in grade school: You can divide p into q as many times as it will go, and the remainder will be smaller than p. There is also a version of the division algorithm for division of polynomials. Formulate such a result.

Combine the Division Algorithm for polynomials with the Fundamental Theorem of Algebra to learn about the factorization of polynomials. Proceed as follows. Suppose that $p(z)$ is a polynomial of degree at least 1 and that r_1 is a root of p, as guaranteed by the Fundamental Theorem of Algebra. Divide $p(z)$ by $z - r_1$. Then there will be some quotient $p_1(z)$ and some remainder $s(z)$. Show that in fact s must be zero. So

$$p(z) = (z - r_1) \cdot p_1(z).$$

Now if the degree of p_1 is at least one, then we may again apply the Fundamental Theorem of Algebra to find a root r_2. Divide $p_1(z)$ by $z - r_2$. The result, as before (using the division algorithm for polynomials) is that

$$p_1(z) = (z - r_2) \cdot p_2(z)$$

for some polynomial p_2, hence

$$p(z) = (z - r_1) \cdot (z - r_2) \cdot p_2(z).$$

Continuing in this manner, we ultimately come up with a factorization

$$p(z) = c_0 \cdot (z - r_1) \cdot (z - r_2) \cdots (z - r_k),$$

where, c_0 is a constant and k is the degree of the original polynomial p.

49. Prove parts (2), (3), (4) of Proposition 5.6.7.

50. Prove parts (1), (2), (3) of Proposition 5.6.8.

51. Refer to Exercise 5.48 for a discussion of the Division Algorithm. Use the Division Algorithm to generate a method for finding the greatest common divisor of two positive integers m and n.

* **52.** Prove that there is a real number $m > 0$ with the following property: Let $a_1, a_2, \ldots, a_k$ be complex numbers. Then there is a subcollection $a_{j_1}, a_{j_2}, \ldots, a_{j_p}$ such that

$$\left| a_{j_1} + a_{j_2} + \cdots + a_{j_p} \right| \geq m \cdot \left[|a_1| + |a_2| + \cdots |a_k| \right] .$$

  [**Hint:** First consider the case $k = 2$.]

  53. Refer to Exercises 5.24, 5.25, 5.26. Produce an explicit complex number z such that $e^z = i$. Produce an explicit complex number w such that $e^w = -i$.

6

More on the Real Number System

6.0 Introductory Remark

The real numbers are a profound and complex world. Section 5.4 introduced you to the real numbers, but did not show you any of its truly deep properties.

In this brief chapter, we begin to explore the real numbers and establish some of its more remarkable aspects. This will be a real mathematical adventure, and you should prepare to enjoy it.

6.1 Sequences

If x is a real number, then the *absolute value* of x, denoted $|x|$, is the distance of x to 0. In other words,

$$|x| = \begin{cases} x & \text{if} & x > 0 \\ 0 & \text{if} & x = 0 \\ -x & \text{if} & x < 0. \end{cases}$$

Fundamental to our study of the deeper properties of the real numbers is the *triangle inequality*: If x, y are real numbers, then

$$|x + y| \le |x| + |y|. \tag{$*$}$$

In fact the standard triangle inequality $(*)$ entails other inequalities that are also useful. Let $x = a + b$ and $y = -b$. Then $(*)$ implies

$$|(a + b) - b| \le |a + b| + |b|$$

hence

$$|a| - |b| \le |a + b|. \tag{$**$}$$

A *sequence* in $\mathbb{R}$ is a function $\phi : \mathbb{N} \to \mathbb{R}$. We denote the elements of the sequence by $\phi(1), \phi(2), \ldots$ For example,

$$\phi(j) = j^2 + 1$$

DOI: 10.1201/9781003214564-6

is a sequence. It is often useful to write out the elements of the sequence in order: $2, 5, 10, 17, \ldots$. We frequently denote the elements of a sequence by the more convenient notation $\phi_1, \phi_2, \phi_3, \ldots$ (rather than think of the sequence as a function). The principal property of a sequence is whether or not it *converges*.

We say that a sequence $\{a_j\} = \{a_1, a_2, \ldots\}$ *converges* to a number α if, for every $\epsilon > 0$, there is a positive integer K such that $j > K$ implies that $|a_j - \alpha| < \epsilon$. What we have enunciated is a quantitative, rigorous way of asserting that the numbers a_j become closer and closer, and *stay* close, to α (within any desired distance ϵ).

EXAMPLE 6.1.1 Consider the sequence $\phi(j) = (-1)^j$, or

$$-1, 1, -1, 1, \ldots .$$

This sequence does *not* converge. Intuitively, the assertion is clear; because the numbers in the sequence do not get close and stay close to any fixed value α. To verify this claim rigorously, we suppose (seeking a contradiction) that in fact the sequence *does* converge to some number α. Let $\epsilon = 1/2$. Then, by the definition of convergence, there is a positive integer K such that if $j > K$, then $|\phi(j) - \alpha| < \epsilon = 1/2$. Choose $j > K$ so that $\phi(j) = 1$, that is to say, choose j even and greater than K. Then $\phi(j+1) = -1$. As a result,

$$
\begin{aligned}
2 = |1 - (-1)| \;&=\; |\phi(j) - \phi(j+1)| = \left| (\phi(j) - \alpha) + (\alpha - \phi(j+1)) \right| \\
&\leq\; |\phi(j) - \alpha| + |\alpha - \phi(j+1)| < \frac{1}{2} + \frac{1}{2} = 1 .
\end{aligned}
$$

We have derived the untenable assertion that $2 < 1$. This contradiction must mean that our assumption is false: the limit number α cannot exist. So the sequence has no limit.   ◇

EXAMPLE 6.1.2 Consider the sequence $\phi(j) = (-1)^j/j$. Intuitively, this sequence converges. For the elements of the sequence seem to be getting smaller and smaller in absolute value, and indeed seem to tend to zero. Let us prove that this actually is the case.

Let $\epsilon > 0$. There is a natural number K so large that $1/K < \epsilon$ (this is the Archimedian property of the natural numbers). If $j > K$, then

$$|\phi(j) - 0| = |\phi(j)| = \frac{1}{j} < \frac{1}{K} < \epsilon ,$$

as was to be proved. So the sequence $\phi(j)$ converges to 0.   ◇

Let $\{a_j\}$ be a sequence. A *subsequence* of $\{a_j\}$ is a sequence $\{b_k\}$ whose elements come from the sequence $\{a_j\}$, in order. We usually denote the subsequence by $\{a_{j_k}\}$.

EXAMPLE 6.1.3 Let

$$a_1 = 1 \ , \ a_2 = 4 \ , \ a_3 = 9 \ , \ a_4 = 16 \ , \ a_5 = 25 \ , \ a_6 = 36 \ , \ \ldots \ , \ a_j = j^2 \ , \ \ldots$$

Then

$$a_{j_1} = 4 \ , \ a_{j_2} = 9 \ , \ a_{j_3} = 36 \ , \ a_{j_4} = 81 \ , \ \ldots$$

is a subsequence. Of course a given sequence will have many different subsequences. ◇

6.2 Open Sets and Closed Sets

Some of the most commonly used subsets of the real numbers are intervals. The four types of intervals are these:

open	$(a, b) = \{x \in \mathbb{R} : a < x < b\}$.
closed	$[a, b] = \{x \in \mathbb{R} : a \le x \le b\}$.
half-open	$[a, b) = \{x \in \mathbb{R} : a \le x < b\}$.
half-open	$(a, b] = \{x \in \mathbb{R} : a < x \le b\}$.

A set $\mathcal{O} \subset \mathbb{R}$ is said to be *open* if, for any $x \in \mathcal{O}$, there is an $\epsilon > 0$ such that $(x - \epsilon, x + \epsilon) \subset \mathcal{O}$. A set $\mathcal{E} \subset \mathbb{R}$ is said to be *closed* if $^c\mathcal{E} \equiv \mathbb{R} \setminus \mathcal{E}$ is open.

A common mistake that students make is to suppose that if a set is not open then it is closed. This is incorrect. The set $[0, 1) = \{x \in \mathbb{R} : 0 \le x < 1\}$ is neither open nor closed.

EXAMPLE 6.2.1 Let $\mathcal{O} = \{x \in \mathbb{R} : x^2 < 1\}$. Then $\mathcal{O}$ is an open set. To see this, let $x \in \mathcal{O}$. Then certainly $|x| < 1$. Let $\epsilon = 1 - |x|$. Then we claim that $(x - \epsilon, x + \epsilon) \subset \mathcal{O}$. For if $t \in (x - \epsilon, x + \epsilon)$, then

$$|t| < |x| + |t - x| < |x| + \epsilon = |x| + (1 - |x|) = 1 \ .$$

Therefore $t^2 < 1$ and $t \in \mathcal{O}$. Thus $(x - \epsilon, x + \epsilon) \subset \mathcal{O}$. As a result, $\mathcal{O}$ is open. ◇

EXAMPLE 6.2.2 Let $\mathcal{E} = \{x \in \mathbb{R} : x^2 \le 1\}$. Then $\mathcal{E}$ is a closed set. To see this, we consider

$$^c\mathcal{E} = \{x \in \mathbb{R} : x < -1 \text{ or } x > 1\} \ .$$

Now let $x \in {}^c\mathcal{E}$. In case $x > 1$, then let $\epsilon = x - 1$. We claim that $(x - \epsilon, x + \epsilon) \subset {}^c\mathcal{E}$. For if $t \in (x - \epsilon, x + \epsilon)$, then

$$t \ge x - |x - t| > x - \epsilon - x - (x - 1) = 1 \ .$$

Thus $t \in {}^c\mathcal{E}$ so $(x - \epsilon, x + \epsilon) \subset {}^c\mathcal{E}$. A similar argument shows that in case $x < -1$ and $\epsilon = (-1) - x$, then $(x - \epsilon, x + \epsilon) \subset {}^c\mathcal{E}$. As a result, $^c\mathcal{E}$ is open; so $\mathcal{E}$ is closed. ◇

Proposition 6.2.3 *Let $\mathcal{E} \subset \mathbb{R}$ be a closed set. Let $\{a_j\}$ be a sequence of points in E. If the sequence $\{a_j\}$ converges to a point $\alpha \in \mathbb{R}$, then $\alpha \in \mathcal{E}$. It is common to say that the set $\mathcal{E}$ contains all its limit points (or accumulation points).*

Proof: Suppose that the assertion is false. Then there is a sequence $\{a_j\} \subset \mathcal{E}$ that converges to a point α, and $\alpha \in {}^c\mathcal{E}$. But $\mathcal{E}$ is closed, so ${}^c\mathcal{E}$ is open. Therefore there is a number $\epsilon > 0$ such that $(\alpha - \epsilon, \alpha + \epsilon) \subset {}^c\mathcal{E}$. But then $a_j \notin (\alpha - \epsilon, \alpha + \epsilon)$ for every j. As a result, $|a_j - \alpha| \geq \epsilon$ for every j. Therefore it cannot be that $\{a_j\}$ converges to α, and that is a contradiction. $\qquad\square$

Proposition 6.2.4 *Suppose that $\mathcal{E} \subset \mathbb{R}$ contains all its limit points. Then E is closed.*

Proof: Suppose not. Then ${}^c\mathcal{E}$ is not open. So there is a point $y \in {}^c\mathcal{E}$ such that, for each integer $j > 0$, the interval $(y - 1/j, y + 1/j) \not\subset {}^c\mathcal{E}$. That means that, for each j, there is a point $e_j \in \mathcal{E}$ that lies in $(y - 1/j, y + 1/j)$. But then $\{e_j\} \subset \mathcal{E}$ and this sequence converges to y. Since $\mathcal{E}$ contains all its limit points, we conclude that $y \in \mathcal{E}$. That is a contradiction. $\qquad\square$

6.3   Compact Sets

A set $S \subset \mathbb{R}$ is said to be *compact* if, whenever $\{s_j\}_{j=1}^\infty$ is a sequence of points in S, then there is a subsequence $\{s_{j_k}\}$ that is convergent to a point of S.

Proposition 6.3.1 *A compact set is closed and bounded.*

Proof: Let K be the compact set. If K is not bounded, then there is an element $x_1 \in K$ such that $|x_1| > 1$. Since K is unbounded, there then exists an element $x_2 \in K$ such that $|x_2| > |x_1| + 1$. Suppose, inductively, that $x_1, x_2, \ldots, x_j$ have been chosen. Then, since K is unbounded, there is an $x_{j+1} \in K$ such that $|x_{j+1}| > |x_j| + j$. It is then clear that the sequence $\{x_j\}$ contains no convergent subsequence. That contradicts the definition of compactness.

If K is not closed, then there is a sequence $\{x_j\} \subset K$ that converges to a point α that does not lie in K. But then every subsequence of $\{x_j\}$ also converges to $\alpha \notin K$. Thus K is not compact. $\qquad\square$

Proposition 6.3.2 *If a subset $E \subset \mathbb{R}$ is both closed and bounded, then it is compact.*

Proof: Let $\{x_j\}$ be a sequence in E. Since E is bounded, it therefore lies in some interval $[-R, R]$. Let

$$S = \{x \in [-R, R] : \exists \text{ infinitely many } x_j$$

$$\text{such that } x_j \geq x\}.$$

Then S is a bounded set, for $|x| \leq R$ for every $x \in S$. Now let α be the least upper bound of S. Then α is finite, indeed $\alpha \leq R$.

We claim that there is a subsequence $\{x_{j_k}\}$ that converges to α. Let $\epsilon > 0$. By the definition of "least upper bound," there must be infinitely many of the x_j between $\alpha - \epsilon$ and $\alpha + \epsilon$, otherwise we chose the least upper bound incorrectly. This assertion is true for every $\epsilon > 0$. Thus

- We may choose x_{j_1} in $(\alpha - 1, \alpha + 1)$;

- We may choose x_{j_2} so that $j_2 > j_1$ and $x_{j_2} \in (\alpha - 1/2, \alpha + 1/2)$;

- We may choose x_{j_3} so that $j_3 > j_2$ and $x_{j_3} \in (\alpha - 1/3, \alpha + 1/3)$;

and so forth. By design, the subsequence $\{x_{j_k}\}$ converges to α. Since E is closed, we may conclude that $\alpha \in E$. Thus E is compact. $\square$

The last two propositions taken together are known as the Heine–Borel theorem: A subset of $\mathbb{R}$ is compact if and only if it is closed and bounded.

We conclude with an important result about the intersection of a nested sequence of sets. We will make decisive use of it in the next section. First, we give an example.

EXAMPLE 6.3.3 For $j = 1, 2, 3, \ldots$, let

$$U_j = \left\{x \in \mathbb{R} : 0 < x < \frac{1}{j}\right\}.$$

Then $U_1 \supseteq U_2 \supseteq U_3 \supseteq \cdots$. And each U_j is uncountable. One's intuition might suggest that $\mathcal{U} \equiv \cap_j U_j$ will certainly have points in it. But that is not the case. In fact, if $x > 0$, then there is a positive integer j so large that $1/j < x$. But then $x \notin U_j$, so certainly $x \notin \mathcal{U}$. If $x \leq 0$, then x does not lie in any U_j so $x \notin \mathcal{U}$. Thus $\mathcal{U}$ is empty. $\diamond$

The situation is more favorable for nested sequences of compact sets:

Theorem 6.3.4 Let $E_1 \supseteq E_2 \supseteq E_3 \supseteq \cdots$ be compact, non-empty sets. Then $\mathcal{E} \equiv \cap_j E_j \neq \emptyset$.

Proof: Let $e_1 \in E_1$, $e_2 \in E_2$, and so forth. The sequence $\{e_j\}$ lies in E_1, which is compact. So there is a subsequence $\{e_{j_k}\}$ that converges to a point $e \in E_1$. But all of the terms of this subsequence for $k \geq 2$ also lie in E_2.

Since E_2 is closed, we may conclude that the limit point e is also in E_2. Continuing, we find that $e \in E_j$ for every j. As a result, $e \in \cap_j E_j \equiv \mathcal{E}$. So $\mathcal{E} \neq \emptyset$. $\square$

It should be noted that the set $\mathcal{E}$ in the theorem is itself compact. For it is the intersection of closed sets, so it is closed. And it is the intersection of bounded sets, so it is bounded. By Heine-Borel, it is compact.

6.4 The Cantor Set

We now use the ideas developed in the first three sections to demonstrate the existence of a remarkable set of real numbers that is known as the "Cantor ternary set." Note, as usual, that a careful and rigorous understanding of the real numbers is necessary in order to effect this construction.

6.4.1 Construction of a remarkable compact set

Let $I = [0,1]$, the unit interval in the real line. Define a sequence of nested compact sets as follows:

- $I_0 = [0,1]$;

- $I_1 = [0,1/3] \cup [2/3,1]$;

- $I_2 = [0,1/9] \cup [2/9,1/3] \cup [2/3,7/9] \cup [8/9,1]$;

- $\ldots$

I_0

FIGURE 6.1
First step in the construction of the Cantor set.

The construction continues. Each set I_{j+1} is constructed by removing the middle open third from each closed interval in the set I_j. Refer to Figures 6.1–6.3 for the sets I_j.

Obviously $I_0 \supseteq I_1 \supseteq I_2 \supseteq \cdots$ and every I_j is non-empty. Also each I_j is closed and bounded, hence compact. Therefore $C = \cap_j I_j$ is a non-empty compact set. We call it *the Cantor ternary set*.

Now we prove a sequence of propositions to establish some of the fundamental properties of the Cantor ternary set.

$$I_1$$

FIGURE 6.2
Second step in the construction of the Cantor set.

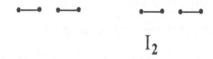

$$I_2$$

FIGURE 6.3
Third step in the construction of the Cantor set.

Proposition 6.4.1 *Let* $S = [0,1] \setminus C$. *Then* S *is a union of intervals with total length 1.*

Of course the unit interval $[0,1]$ itself has length 1. The proposition establishes that the complement of the Cantor set in the unit interval has length 1. These assertions together suggest that the Cantor set has length 0, so that it is "small" in some sense. In the subject of measure theory, one makes these assertions precise.

Proof of the Proposition: Of course

$$
\begin{aligned}
[0,1] \setminus C &= [0,1] \cap {}^c C \\
&= [0,1] \cap {}^c\left(\bigcap_j I_j\right) \\
&= [0,1] \cap \left(\bigcup_j {}^c I_j\right) \\
&= \bigcup_j \left([0,1] \cap {}^c I_j\right).
\end{aligned}
$$

And each of the sets $[0,1] \cap {}^c I_j$ is a union of intervals. Thus we see explicitly that the complement of the Cantor set is a union of intervals. More is true: these complements form an increasing union. So it is easy to keep track of all the intervals and to sum up their lengths:

- First, there is a single interval of length $1/3$.

- Second, there are two intervals of length $1/9$.

- Third, there are four intervals of length $1/27$.

- ...

Thus we may add up the lengths of all the intervals in the complement of C:

$$\text{length}(^cC) = \sum_{j=1}^{\infty} \frac{2^{j-1}}{3^j} = \frac{1}{3} \cdot \sum_{j=0}^{\infty} \left(\frac{2}{3}\right)^j = \frac{1}{3} \cdot \frac{1}{1-2/3} = 1 \,. \qquad \square$$

So, in the sense of length, the Cantor set is small. Now we shall show that, in the sense of cardinality, the Cantor set is large.

Proposition 6.4.2 *The Cantor set is uncountable.*

Proof: We assign a binary "address" to each element of the Cantor set as follows. The address will be a sequence of 0s and 1s. Fix an element x of the Cantor set. Examine Figure 6.2. The first entry of the address is a 0 if x is in the left component of I_1; otherwise the first entry is a 1. Now look at Figure 6.3. An open segment is removed from the segment in Figure 6.2 in which x resides. Thus x is either in the resulting left component or in the resulting right component. As a result, the second address entry is either 0 or 1, respectively. Continue in this fashion, assigning the jth address entry to be 0 or 1 according as x lies in either the left or right component of the divided interval in I_j.

We see that we may assign to each element of the Cantor set a sequence of 0s and 1s. Conversely, each sequence of 0s and 1s determines an element of the Cantor set using the same logic. We learned in Section 4.5 that the set of all sequences of 0s and 1s is uncountable. Hence the Cantor set is uncountable. $\square$

We conclude this discussion, and indeed the chapter, by constructing a counterintuitive function. Called the "Cantor–Lebesgue function," it has the property that it is continuous, its derivative is 0 at most points (i.e., on the complement of the Cantor set), yet the function is not constant. A graph of the function is exhibited in Figure 6.4.

Proposition 6.4.3 *There exists a continuous function c with domain the interval $[0,1]$ and range the interval $[0,1]$ such that*

(1) *The function takes the values $c(0) = 0$ and $c(1) = 1$;*

(2) *The function c is monotone increasing;*

(3) *On the complement of the Cantor set, c is locally constant.*

Proof of the Proposition: Define the Cantor–Lebesgue function in this way:

$$c(x) = \begin{cases} \sum_{j=1}^{\infty} i_j \cdot (2^{-j})^{i_j} & \text{if } x \in C \text{ and } x \text{ has address } \{i_j\} \\ \sup\{c(t) : t \in C \text{ and } t \le x\} & \text{if } x \notin C. \end{cases}$$

Here sup is the abbreviation for "supremum," which is the same thing as "least upper bound." In the first line of the definition, on the right, we use

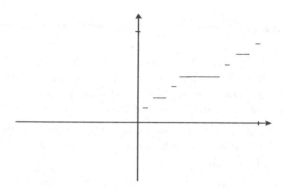

FIGURE 6.4
The Cantor–Lebesgue function.

the "address" (the sequence of 0s and 1s) that was our device for seeing that the Cantor set is uncountable.

Notice that we first define the Cantor–Lebesgue function for x in the Cantor set, then we define the function at points not in the Cantor set *in terms of* our earlier definition at points of the Cantor set.

This definition is rather technical. It can also be described in words. On the single interval that is removed in the construction of I_1, we declare the Cantor–Lebesgue function to have value $1/2$. On the pair of intervals removed in the construction of I_2, we declare the Cantor–Lebesgue function to have values $1/4$ (on the left) and $3/4$ (on the right). On the four intervals removed in the construction of I_3, we declare the Cantor–Lebesgue function to have values $1/8, 3/8, 5/8, 7/8$. And so on. The values of the Cantor–Lebesgue function at the intermediate points of the Cantor set are obtained by interpolation.

It is clear from our rigorous definition of the Cantor–Lebesgue function that it is monotone increasing. For if $x_1 < x_2$ are elements of the Cantor set then the address of x_2 will have more 1's in it than the address of x_1, hence the sum defining the value of c will have more positive terms. Likewise, if $x_1 < x_2$ are *not* elements of the Cantor set, and have an element of the Cantor set between them, then the sup defining $c(x_2)$ will be a supremum over *more* numbers than the supremum defining $c(x_1)$, and hence will be larger. If there are *no* elements of the Cantor set between x_1 and x_2 then $c(x_1) = c(x_2)$. The verification of monotonicity for one point *in* the Cantor set and one point *not in* the Cantor set is immediate from the definitions.

The continuity of the Cantor–Lebesgue function follows from similar, but more technical, considerations. It is the proper subject matter of a course in real analysis. We refer the reader to [RUD] for the full details of this matter. □

See https://mathworld.wolfram.com/CantorSet.html for more about the Cantor set.

Exercises

1. Let $0 < \eta < 1$. Modify the construction of the Cantor ternary function to obtain a set $C_\eta \subset [0,1]$ with the property that $[0,1] \setminus C_\eta$ is a union of intervals the sum of whose lengths is η.

2. Verify that if $x_1 < x_2$ are elements of the Cantor set, then there is a number t that is strictly between x_1 and x_2 and such that t does not lie in the Cantor set.

3. Prove that the collection of endpoints of the removed open intervals in the Cantor set construction is countable. Such points have addresses with only finitely many 1s. Thus most elements of the Cantor set are not endpoints, but are points with addresses that have infinitely many nonzero entries.

* 4. A subset of $\mathbb{R}$ is said to be *perfect* if it is closed and bounded and every element of the set is a limit point of the set. Prove that the Cantor set is perfect. [*Remark:* It is an advanced theorem that a perfect set must perforce be uncountable. This gives another proof of the uncountability of the Cantor set.]

5. Let $S \subset \mathbb{R}$. A point $s \in S$ is said to be an *interior* point if there exists a number $\epsilon > 0$ such that $(s - \epsilon, s + \epsilon) \subset S$. The interior of S is just the collection of all interior points. Prove that the interior of S is an open set.

6. Let $S \subset \mathbb{R}$. A point $s \in \mathbb{R}$ is said to be a *boundary point* of S if, for each $\epsilon > 0$, we have both $(s-\epsilon, s+\epsilon) \cap S \neq \emptyset$ and $(s-\epsilon, s+\epsilon) \cap {}^c S \neq \emptyset$. The boundary of S is just the collection of all boundary points. Prove that the boundary of S is closed.

7. Refer to the last two exercises for terminology. Let $S \subset \mathbb{R}$ be any closed subset. Prove that S is the union of its interior and its boundary.

8. Let $\{x_j\}$ be a sequence of real numbers. Prove that the sequence $\{x_j\}$ converges to α if and only if every subsequence has itself a subsequence that converges to α.

9. A set $S \subset \mathbb{R}$ is said to be *discrete* if, for each point $s \in S$, there is a number $\epsilon > 0$ such that $(s - \epsilon, s + \epsilon) \cap S = \{s\}$. Is the set $\mathbb{Q}$ of rational numbers discrete? Is the Cantor set discrete? Is the set $\mathbb{Z}$ of integers discrete? Define

$$T = \{1, 1/2, 1/3, \dots\}.$$

Is T discrete?

* **10.** Refer to Exercise 6.4 for the definition of perfect set. Prove that every compact set K is the union of a perfect set and a countable set. [**Hint:** Define x to be a *consolidation point* of K if each open interval about x contains uncountably many points of K. Let the perfect set consist of the set of consolidation points.]

11. Let $s_1 \le s_2 \le s_3 \cdots$ be real numbers. Prove that either the sequence $\{s_j\}$ converges or else it is unbounded.

12. A set $W \subset \mathbb{R}$ is said to be *dense* in $\mathbb{R}$ if, for each $x \in \mathbb{R}$ and each $\epsilon > 0$, the interval $(x - \epsilon, x + \epsilon)$ contains an element of W. Prove that the set of irrational numbers is dense in $\mathbb{R}$. Give an example of a sequence that is dense in $\mathbb{R}$.

13. Let $S \subset \mathbb{R}$ be any set. Let L be the set of limit points of S. Prove that L is closed.

14. Let $S \subset \mathbb{R}$ be any set. The *closure* of S is defined to be the union of S with all its limit points (see Exercise 6.13). Prove that the closure of S is a closed set. What is the closure of $\mathbb{Q}$? What is the closure of $\mathbb{Z}$?

15. Refer to Exercises 6.5, 6.6, and 6.14 for terminology. Let $S \subset \mathbb{R}$ be any closed set, let $\mathcal{O}$ be its interior, and let $\mathcal{C}$ be its closure. Prove that $\mathcal{C} \setminus \mathcal{O}$ is just the boundary of S.

16. Let $U \subset \mathbb{R}$ be any open set. Prove that U can be written as the countable disjoint union of open intervals. [**Hint:** The intervals will be the equivalence classes under a suitable equivalence relation.]

17. Refer to the last exercise. Let $E \subset \mathbb{R}$ be any closed set. Prove that there is a continuous function ϕ with domain $\mathbb{R}$ such that $\phi \ge 0$ and $Z = \{x \in \mathbb{R} : \phi(x) = 0\} = E$.

* **18.** Let $S \subset \mathbb{R}$ be any set. We say that S is *disconnected* if there are open sets U and V with $U \cap V = \emptyset$ such that $U \cap S \ne \emptyset$, $V \cap S \ne \emptyset$, and

$$S = [U \cap S] \cup [V \cap S].$$

If S is not disconnected, then we say that S is *connected*. Prove that $\mathbb{R}$ is connected. Prove that any interval is connected. Prove that $\mathbb{Q}$ is disconnected. Prove that the Cantor set is disconnected.

19. Refer to Exercise 6.14 for terminology. Let $\{x_j\}$ be any sequence in $\mathbb{R}$. Let X be the closure of $\{x_j\}$. How many points can $X \setminus \{x_j\}$ have in it? One? Finitely many? Infinitely many? Uncountably many?

20. Refer to Exercise 6.5 for terminology. What is the interior of $\mathbb{Q}$? What is the interior of $\mathbb{R}$? What is the interior of the Cantor set?

21. Let $\mathcal{O}_1, \mathcal{O}_2, \ldots$ be open sets. Define $\mathcal{O} = \cup_j \mathcal{O}_j$. Is $\mathcal{O}$ open? Define $\mathcal{T} = \cap_j \mathcal{O}_j$. Is T open?

Let $\mathcal{E}_1, \mathcal{E}_2, \ldots$ be closed sets. Define $\mathcal{E} = \cup_j \mathcal{E}_j$. Is $\mathcal{E}$ closed? Define $\mathcal{F} = \cap_j \mathcal{E}_j$. Is $\mathcal{F}$ closed?

22. Refer to the last exercise. For the assertions that are false, show that if one restricts attention to finite collections of sets, then they become true.

23. Give a characterization of the open subsets of $\mathbb{R}$ using the language of sequences.

24. Give a characterization of the closed subsets of $\mathbb{R}$ using the language of neighborhoods.

25. Let S be an uncountable subset of $\mathbb{R}$. Show that S must have a limit point. That is, there is a point $s \in \mathbb{R}$ and a sequence $\{s_j\} \subset S$ with $s_j \neq s$ for all j such that $\lim_{j \to \infty} s_j = s$. Prove that if S has only countably many points, then the conclusion fails.

26. Let $\mathcal{O}$ be an open subset of $\mathbb{R}$. Prove that $\mathcal{O}$ can be written as the countable increasing union of closed sets.

27. Let E be a closed subset of $\mathbb{R}$. Prove that E can be written as the countable decreasing intersection of open sets.

28. Explain why the words "closed" and "open" may not be switched in Exercise 26. Explain why the words "closed" and "open" may not be switched in Exercise 27.

* 29. Define "open set" in $\mathbb{R}^2$. Prove that the product of open sets in $\mathbb{R}^1$ is open.

* 30. Define "closed set" in $\mathbb{R}^2$. Prove that the product of closed sets in $\mathbb{R}^1$ is closed.

7

Elementary Number Theory

Number theory is one of the oldest part of mathematics. It was Euclid who proved that there are infinitely many prime numbers, and this was one of the very first results of number theory. Euclid also deduced the Fundamental Theorem of Arithmetic—that every positive integer has a unique factorization into primes.

Today number theory is a flourishing part of the enterprise. Andrew Wiles's solution of Fermat's Last Theorem [SIN] is certainly one of the more dramatic events. But number theory is finding applications in many parts of our lives—especially cryptography. In this chapter, we introduce some of the number-theoretic ideas that are useful in cryptography, and in the next chapter we show how to use them.

7.1 Prime Numbers

As noted earlier in the book, a prime number is a positive integer, greater than 1, which has no divisors except for 1 and itself. The first several primes are

$$2, 3, 5, 7, 11, 13, 17, 19, 23, 29, 31, 37, 41, 43 \, .$$

We know from our earlier work (see Section 2.3) that there are infinitely many prime numbers. Every positive integer can be factored in a unique way as the product of primes.

Going back at least to the time of Carl Friedrich Gauss (1777–1855), mathematicians have been interested in the distribution of primes. If n is a positive integer, then define $\pi(n)$ to equal the number of primes less than or equal to n. We would like to know the size of $\pi(n)$ for n large.

The celebrated Prime Number Theorem, proved in 1896 independently by J. Hadamard (1865–1963) and C. de la Vallée Poussin (1866–1962), states that $\pi(n)$ is asymptotically of size $n/\log n$ as $n \to \infty$. In other words,

$$\lim_{n \to \infty} \frac{\pi(n)}{n/\log n} = 1 \, .$$

Today there are several proofs of this result. The traditional proof uses complex analysis and, particularly the Riemann zeta function. It is too advanced

DOI: 10.1201/9781003214564-7

167

for presentation in this book, so we omit it. It was considered a remarkable event when, in 1949, Paul Erdős (1913–1996) [ERD] and Atle Selberg (1917–2007) [SEL] found an "elementary" proof of the Prime Number Theorem. Here "elementary" means that the proof uses no complex analysis; only elementary number theory. It is still too difficult to present here.

Paul Erdős (1913–1996)

Paul Erdős came from a Jewish family. Paul's parents had two daughters, aged three and five, who died of scarlet fever just days before Paul was born. This naturally had the effect making Lajos and Anna extremely protective of Paul. He would be introduced to mathematics by his parents, themselves both teachers of mathematics.

The year 1920 was not all bad for Paul, for his father Lajos returned home from captivity in Siberia. He had learnt English to pass the long hours in captivity but, having no English teacher, did not know how to pronounce the words. He now set about teaching Paul to speak English, but this gave Paul a strange English accent.

Erdős, as the winner of a national examination, was allowed to enter the university in 1930. He studied for his doctorate at the University Pázmány Péter in Budapest. Awarded a doctorate in 1934, he took up a postdoctoral fellowship at Manchester, essentially being forced to leave Hungary because he was Jewish.

The situation in Hungary by the late 1930s clearly made it impossible for someone of Jewish origins to return. However he did visit Budapest three times a year during his tenure of the Manchester fellowship. Within weeks, Erdős was on his way to the USA where he took up a fellowship at Princeton.

When Erdős was 18 years old, he found an elementary proof of the assertion that there is always a prime between n and $2n$ for $n \geq 2$.

Another result on prime numbers associated with Erdős is the Prime Number Theorem, namely, the number of primes $\leq n$ tends to ∞ as $n/\log n$. In 1949, Erdős and Atle Selberg found an elementary proof. Selberg and Erdős agreed to publish their work in back-to-back papers in the same journal, explaining the work each had done and sharing the credit. But at the last minute Selberg raced ahead with his proof and published first. Erdős was not much concerned with the competitive aspect of mathematics and was philosophical about the episode.

Ulam left Madison in 1943 to join other mathematicians and physicists at Los Alamos in New Mexico working on the atomic bomb project. He asked Erdős to join the project but, although he was interested enough to be interviewed, Erdős gave answers to those interviewing him which he must have known were not what they wanted to hear.

In 1943, Erdős worked at Purdue University, taking a part-time appointment. He had heard nothing from his family between 1941 and the time when Budapest was liberated in 1945. It is unlikely that the full extent of the horror was understood by Erdős in the United States at the time. However, in August 1945, Erdős received a telegram giving details of his family.

Near the end of 1948, Erdős was able to return to Hungary for a visit and there he was reunited with his surviving family and friends.

During the early 1950s, senator Joseph R McCarthy whipped up strong feelings against communism in the United States. Erdős began to come under suspicion from authorities who saw imaginary problems everywhere. He spent much of the next ten years in Israel. During the early 1960s, he made numerous requests to be allowed to return to the United States and a visa was finally granted in November 1963. By this time, however, Erdős had become a traveler moving from one university to another, and from the home of one mathematician to another.

Paul Erdős was also one of the most prolific mathematicians in history, with more than 1,500 papers to his name. Erdős won many prizes including the Wolf Prize of $ 50,000 dollars in 1983. He gave away most of the money he earned, donating it to help students or as prizes for solving problems he had posed.

A classical method for identifying prime numbers is the Sieve of Eratosthenes. It is produced in this way. Take a grid of the integers greater than 1:

2	3	4	5	6	7	8	9	10	11
12	13	14	15	16	17	18	19	20	21
22	23	24	25	26	27	28	29	30	31
32	33	34	35	36	37	38	39	40	41
42	43	44	45	46	47	48	49	50	51

The first prime is 2, so we cross out all multiples of 2, of order at least 2, that occur in the grid:

2	3	~~4~~	5	~~6~~	7	~~8~~	9	~~10~~	11
~~12~~	13	~~14~~	15	~~16~~	17	~~18~~	19	~~20~~	21
~~22~~	23	~~24~~	25	~~26~~	27	~~28~~	29	~~30~~	31
~~32~~	33	~~34~~	35	~~36~~	37	~~38~~	39	~~40~~	41
~~42~~	43	~~44~~	45	~~46~~	47	~~48~~	49	~~50~~	51

The second prime is 3, so we next cross out all multiples of 3, of order at least 2, that occur in the grid:

2	3	~~4~~	5	~~6~~	7	~~8~~	~~9~~	~~10~~	11
~~12~~	13	~~14~~	~~15~~	~~16~~	17	~~18~~	19	~~20~~	~~21~~
~~22~~	23	~~24~~	25	~~26~~	~~27~~	~~28~~	29	~~30~~	31
~~32~~	~~33~~	~~34~~	35	~~36~~	37	~~38~~	~~39~~	~~40~~	41
~~42~~	43	~~44~~	~~45~~	~~46~~	47	~~48~~	49	~~50~~	~~51~~

The third prime is 5, so we next cross out all multiples of 5, of order at least 2, that occur in the grid:

2	3	~~4~~	5	~~6~~	7	~~8~~	~~9~~	~~10~~	11
~~12~~	13	~~14~~	~~15~~	~~16~~	17	~~18~~	19	~~20~~	~~21~~
~~22~~	23	~~24~~	~~25~~	~~26~~	~~27~~	~~28~~	29	~~30~~	31
~~32~~	~~33~~	~~34~~	~~35~~	~~36~~	37	~~38~~	~~39~~	~~40~~	41
~~42~~	43	~~44~~	~~45~~	~~46~~	47	~~48~~	49	~~50~~	~~51~~

The fourth prime is 7, so we next cross out all multiples of 7, of order at least 2, that occur in the grid:

2	3	~~4~~	5	~~6~~	7	~~8~~	~~9~~	~~10~~	11
~~12~~	13	~~14~~	~~15~~	~~16~~	17	~~18~~	19	~~20~~	~~21~~
~~22~~	23	~~24~~	~~25~~	~~26~~	~~27~~	~~28~~	29	~~30~~	31
~~32~~	~~33~~	~~34~~	~~35~~	~~36~~	37	~~38~~	~~39~~	~~40~~	41
~~42~~	43	~~44~~	~~45~~	~~46~~	47	~~48~~	~~49~~	~~50~~	~~51~~

The fifth prime is 11, so we next cross out all multiples of 11, of order at least 2, that occur in the grid:

2	3	4̶	5	6̶	7	8̶	9̶	1̶0̶	11
1̶2̶	13	1̶4̶	1̶5̶	1̶6̶	17	1̶8̶	19	2̶0̶	2̶1̶
2̶2̶	23	2̶4̶	2̶5̶	2̶6̶	2̶7̶	2̶8̶	29	3̶0̶	31
3̶2̶	3̶3̶	3̶4̶	3̶5̶	3̶6̶	37	3̶8̶	3̶9̶	4̶0̶	41
4̶2̶	43	4̶4̶	4̶5̶	4̶6̶	47	4̶8̶	4̶9̶	5̶0̶	5̶1̶

Since 11 exceeds the square root of 51, the last number in the grid, we may now cease our efforts. You can see that the numbers that remain (that have not been crossed out) are the primes.

There are a great many properties of the prime numbers that one might explore. We examine one of them right now, just to give you the flavor of the kind of reasoning we can use. Before we begin the proof, let us observe that every positive, odd integer greater than 1 is either of the form $4m + 1$ or of the form $4m - 1$.

Proposition 7.1.1 *There are infinitely many primes of the form $4m - 1$.*

Proof: We begin by noticing that there are plenty of examples of primes of this form:

$$3, 7, 11, 19, 23, 31, 43$$

for instance. We now prove that there are infinitely many of them.

Seeking a contradiction, we suppose that there are only finitely many such primes. Call them $p_1, p_2, \ldots, p_k$. Form the number

$$N = (p_1 \cdot p_2 \cdots \cdot p_k) - 1.$$

Of course $4p_j \nmid N$ (does not divide) for every j. Also notice that not every prime p that divides N is of the form $4m + 1$; for if that were the case then N would be of the form $4m + 1$. And it is not.

Since N is odd, each prime divisor of N is odd. Since not all the (odd) divisors is of the form $4m + 1$, then one of them must be of the form $4m - 1$. Of course $4m - 1$ cannot be one of the p_j. So it must be a *new* prime of the form $4m - 1$. That is a contradiction. □

It is worth considering why this proof does not work for primes of the form $4m + 1$.

A more general theorem of J. Dirichlet (1805–1859), too difficult for us to treat here in any detail, is the following:

Theorem 7.1.2 *Let a and b be integers with greatest common divisor equal to 1. Then there are infinitely many primes of the form $ax + b$.*

7.2   Greatest Common Divisor

If m and n are positive integers, then their *greatest common divisor* is the greatest integer k such that k divides m evenly and k divides n evenly. We commonly write $k|m$ to mean that k divides m. We let $\gcd(m, n)$ denote the greatest common divisor of m and n.

Examples of greatest common divisor are

$$\gcd(6, 27) = 3 \ , \quad \gcd(8, 100) = 4 \ , \quad \gcd(180, 252) = 36 \ , \quad \gcd(18, 35) = 1 \, .$$

A sensible and sometimes convenient way to calculate the greatest common divisor of two integers is to examine their prime factorizations. For example, to determine the greatest common divisor of 180 and 252 we note that

$$180 = 2 \cdot 2 \cdot 3 \cdot 3 \cdot 5 \quad \text{and} \quad 252 = 2 \cdot 2 \cdot 3 \cdot 3 \cdot 7 \, .$$

We see that these integers have the prime factors $2, 2, 3, 3$ in common. So the greatest common divisor is $\gcd(180, 252) = 2 \cdot 2 \cdot 3 \cdot 3 = 36$.

Now we shall use the Division Algorithm to find a more insightful and constructive way to find the greatest common divisor.

Lemma 7.2.1 *For any positive integers a and b, we have*

$$\gcd(a, b) = \gcd(b, a) = \gcd(\pm a, \pm b) = \gcd(a, b - a) = \gcd(a, b + 1) \, .$$

Proof: The proofs of all these equalities are quite similar, so we shall prove only that $\gcd(a, b) = \gcd(a, b-a)$. All we need to do is to note that if k divides both a and b with quotients α and β, then k divides a and $b - a$ with quotients α and $\beta - \alpha$. Thus $\gcd(a, b) \leq \gcd(a, b - a)$. Likewise, if j divides both a and $b - a$ with quotients α and $\beta - \alpha$, then j divides a and b with quotients α and $\beta = (\beta - \alpha) + \alpha$. Thus $\gcd(a, b - a) \leq \gcd(a, b)$. $\qquad\qquad\square$

Lemma 7.2.2 *If a, b, n are positive integers, then $\gcd(a, b) = \gcd(a, b - an)$.*

Proof: Similar to the proof of the preceding lemma. $\qquad\qquad\square$

Definition 7.2.3 Recall that the *Division Algorithm* says this. Let a and b be integers with $b \neq 0$. Then there exist unique integers q and r such that

(a) $0 \leq r < |b|$;

(b) $a = bq + r$.

This last is nothing other than the Division Algorithm that you learned in grade school: If you divide a positive integer a by a positive integer b, then you get a quotient q and a remainder r that is less than b.

We next present the famous Euclidean algorithm for finding the greatest common divisor of two integers:

Algorithm for Finding the Greatest Common Divisor: Let a and b be positive integers. Assume that $a > b > 0$. To calculate $\gcd(a, b)$, perform the following steps:

(1) Use the Division Algorithm to write $a = bq + r$ with $0 \leq r < b$ and $q \in \mathbb{Z}$.

(2) If $r = 0$ then $b|a$ and hence the greatest common divisor is b and the process ceases.

(3) If $r > 0$ then go back to Step (1), replacing a by b and b by r.

(4) This iterative process will ultimately terminate with $r = 0$. When it does so, you can read off that the last value for b is the greatest common divisor.

Proof of the Algorithm: Lemma 7.2.1 and 7.2.2 tell us that $\gcd(a, b) = \gcd(b, r)$. Hence the gcd does not change in Step (3). Since the remainders form a decreasing sequence of non-negative integers, the process ceases. □

EXAMPLE 7.2.4 Let us use the algorithm to calculate the greatest common divisor of 1275 and 2261. To make things very explicit we shall number the steps.

(1) $2261 = 1 \cdot 1275 + 986$.

(2) $r = 986 > 0$.

(3) We go back to Step 1, this time dividing 1275 by 986.

(1) $1275 = 1 \cdot 986 + 289$.

(2) $r = 289 > 0$.

(3) We go back to Step 1, this time dividing 986 by 289.

(1) $986 = 3 \cdot 289 + 119$.

(2) $r = 119 > 0$.

(3) We go back to Step 1, this time dividing 289 by 119.

(1) $289 = 2 \cdot 119 + 51$.

(2) $r = 51 > 0$.

(3) We go back to Step 1, this time dividing 119 by 51.

(1) $119 = 2 \cdot 51 + 17$.

(2) $r = 17 > 0$.

(3) We go back to Step 1, this time dividing 51 by 17.

(1) $51 = 3 \cdot 17 + 0$.

(2) $r = 0$.

(4) The process has terminated. The last value for b is 17, and that is the greatest common divisor of 1275 and 2261.

$\diamond$

We conclude this section with an elegant result that is useful, among other things, in cryptography.

Theorem 7.2.5 *Let $a, b \in \mathbb{Z}$ and set $g = \gcd(a, b)$. Then there exist $x, y \in \mathbb{Z}$ such that*

$$ax + by = g\,.$$

Remark 7.2.6 This result is particularly interesting when a and b have no common factors, so that $g = 1$.

Also note that, if $e = cg$ is a multiple of g, then

$$cax + cby = cg = e$$

so that

$$e = (cx)a + (cy)b\,.$$

So, we can express e in terms of a and b.

Proof: Look ahead to remind yourself how modular arithmetic works.

Observe that $\gcd(a/g, b/g) = 1$, so one can check that the equation

$$\frac{a}{g} \cdot x = 1 \quad \text{mod} \ \frac{b}{g} \qquad\qquad (*)$$

has a solution $x \in \mathbb{Z}$. Multiplying $(*)$ through by g yields $ax = g \mod b$. Hence there exists y such that $b \cdot (-y) = ax - g$. But this says that $ax + by = g$ as was to be proved. $\square$

7.3 Modular Arithmetic

This is a familiar idea, and we have already alluded to it earlier. You know, of course, from your reading of earlier parts of this book that we perform arithmetic mod k by imposing the equivalence relation "$m \sim n$ if $m - n$ is divisible by k" on the integers. The resulting equivalence classes are the elements of arithmetic mod k.

A slightly more informal, but perfectly correct, way to think about the matter is as follows. When we write $n \bmod k$, we mean simply the remainder when n is divided by k. Thus

$$
\begin{aligned}
25 &= 1 \quad \bmod \ 3\,, \\
15 &= 3 \quad \bmod \ 4\,, \\
-13 &= -3 \quad \bmod \ 5 \\
&= 2 \quad \bmod \ 5\,.
\end{aligned}
$$

It is an important fact—which again is most clearly seen using the theory of cosets—that modular arithmetic respects sums and products. That is,

$$
(a + b) \bmod n = a \bmod n + b \bmod n
$$

and

$$
(a \cdot b) \bmod n = \Big[(a \bmod n) \cdot (b \bmod n)\Big] \bmod n\,.
$$

We shall use these facts in a decisive manner below.

7.4 Theorems of Wilson and Chinese Remainder

Wilson's theorem dates back to the 1770s. It was actually proved by Lagrange. We need several preliminary results to get to Wilson's.

Definition 7.4.1 Fix a positive integer n. Let $S \subseteq \mathbb{Z}$ be a set of size n. We call S a *complete set of residues* modulo n if the residues modulo n of the elements of S are distinct. In other words, S contains a representative from each equivalence class in $\mathbb{Z}/n\mathbb{Z}$.

Lemma 7.4.2 *Let S be a complete set of residues modulo n and let $a \in \mathbb{Z}$ with $\gcd(a, n) = 1$. Then $aS \equiv \{as : s \in S\}$ is also a complete set of residues modulo n.*

Proof: If $x, x' \in S$ and $ax = ax' \mod n$, then it is clear that $x = x' \mod n$. Since S is a complete set of residues, this implies that $x = x'$. Thus the elements of aS have distinct residues modulo n. Because the number of elements in aS is n, it follows that aS is a complete set of residues modulo n. □

Proposition 7.4.3 *If $\gcd(a, n) = 1$, then the equation $ax = b \mod n$ has a solution, and that solution is unique modulo n.*

Proof: Let S be a complete set of residues modulo n. Then there is a unique element of S that is congruent to b modulo n. By the preceding lemma, aS is also a complete set of residues modulo n. Hence there is a unique element $ax \in aS$ that is congruent to b modulo n. We conclude that $ax = b \mod n$. □

Proposition 7.4.4 *The equation $ax = b \mod n$ has a solution if and only if $\gcd(a, n)$ divides b.*

Proof: Let $g = \gcd(a, n)$. If there is a unique solution x to the equation $ax = b \mod n$, then $n | (ax - b)$. Since $g | n$ and $g | a$, we conclude that $g | b$.

Conversely, suppose that $g | b$. Then $n | (ax - b)$ if and only if

$$\frac{n}{g} \left| \left(\frac{a}{g} x - \frac{b}{g} \right) \right. .$$

Thus $ax = b \mod n$ has a solution if and only if

$$\frac{a}{g} x = \frac{b}{g} \mod \frac{n}{g}$$

has a solution. Since $\gcd(a/g, n/g) = 1$, Proposition 7.4.3 tells us that this latter equation does have a solution. □

Theorem 7.4.5 (Wilson) *An integer $p > 1$ is a prime if and only if $(p-1)! = -1 \mod p$.*

Proof: The statement is obvious when $p = 2$. So we may as well assume that $p > 2$.

First we assume that p is prime and we shall show that $(p - 1)! = -1 \mod p$. If $a \in \{1, 2, \ldots, p - 1\}$ then of course a is relatively prime to p so that the equation

$$ax = 1 \mod p$$

has a unique solution $a' \in \{1, 2, \ldots, p - 1\}$. If $a = a'$, then $a^2 = 1 \mod p$ so that $p | (a^2 - 1)$ and $(a^2 - 1) = (a - 1)(a + 1)$. Hence $p | (a - 1)$ or $p | (a + 1)$. Thus $a \in \{1, p - 1\}$. We can thus pair up the elements of $\{2, 3, \ldots, p - 2\}$, each with its inverse.

As a result,
$$2 \cdot 3 \cdots (p-2) = 1 \mod p.$$

Multiplying both sides by $p-1$ then proves that $(p-1)! = -1 \mod p$.

For the converse, assume that $(p-1)! = -1 \mod p$ and we shall prove that p must be prime. Suppose not. Then $p \geq 4$ is a composite number. Let ℓ be a prime divisor of p. Then $\ell < p$ so that $\ell | (p-1)!$. Also, by our hypothesis,

$$\ell | ((p-1)! + 1).$$

This is a contradiction because a prime cannot both divide a number a and also $a+1$ (for this would imply that p divides 1, which is impossible). $\square$

Next we turn to the famous Chinese Remainder Theorem. Put simply, the Chinese Remainder Theorem gives conditions under which a system of linear equations is guaranteed to have a solution. The result is, however, remarkable because it says that we can specify more than one residue at a time.

Theorem 7.4.6 *Let $a, b \in \mathbb{Z}$. Let $m, n \in \mathbb{N}$ be such that $gcd(m,n) = 1$. Then there exists an $x \in \mathbb{Z}$ such that*

$$x = a \mod m$$
$$x = b \mod n.$$

Furthermore, x is unique modulo mn.

Proof: If we can solve for t in the equation

$$a + tm = b \mod n, \qquad (*)$$

then $x = a + tm$ will do the job.

To see that we can solve $(*)$, subtract a from both sides and use Proposition 7.4.3 together with the hypothesis that $gcd(m,n) = 1$.

For uniqueness, suppose that both x and y satisfy the two specified congruences. Then $z = x - y$ satisfies $z = 0 \mod m$ and $z = 0 \mod n$. Thus $m|z$ and $n|z$. Since $gcd(m,n) = 1$, we may conclude that $mn|z$. Hence $x = y \mod mn$. $\square$

Remark 7.4.7 We have stated and proved the Chinese Remainder Theorem for two congruences. But in fact the result is true for any finite number of congruences.

7.5   The Euler–Fermat Theorem

Let a and b be two (positive) integers. We say that a and b are relatively prime if they have no common prime factors. For example,

$$72 = 2^3 \cdot 3^2$$

$$175 = 5^2 \cdot 7$$

hence 72 and 175 are relatively prime.

If n is an integer, let $\mathcal{P}(n)$ be the set of integers less than n that are relatively prime to it. Let $\varphi(n)$ be the number of elements in $\mathcal{P}(n)$.

Theorem: If n is a positive integer and k is relatively prime to n, then

$$k^{\varphi(n)} = 1 \bmod n.$$

Proof: The proof of this result is easy. For the collection $\mathcal{P}(n)$ of numbers relatively prime to n forms a group under multiplication.[1] That is, if a is relatively prime to n and b is relatively prime to n, then logic dictates that $a \cdot b$ is relatively prime to n. Now it is a fundamental fact—see Section 11.1—that if a group has m elements and g is an element of the group, then g^m is the group identity. Thus any element of the group, raised to the power $\varphi(n)$ (the number of elements in the group), will equal 1 modulo n. $\square$

For later use, it is worth noting that if p, q are prime numbers and $n = p \cdot q$, then $\varphi(n) = (p - 1) \cdot (q - 1)$. The reason is that the only numbers less than or equal to n that are not relatively prime to n are $p, 2p, 3p, \ldots, q \cdot p$ and $q, 2q, 3q, \cdots, (p - 1)q$. There are q numbers in the first list and $p - 1$ numbers in the second list. The set $\mathcal{P}(n)$ of numbers relatively prime to n is the complement of these two lists, and it therefore has

$$pq - q - (p - 1) = pq - q - p + 1 = (p - 1) \cdot (q - 1) \equiv \varphi(n)$$

elements.

---

[1] We shall study groups in detail in Section 11.1. For now you should think of a group as a set G equipped with a binary operation "$\cdot$". The axioms are **(a)** if $g, h \in G$, then $g \cdot h \in G$, **(b)** the binary operation is associative, **(c)** there is an element $e \in G$ such that, if $g \in G$, then $e \cdot g = g \cdot e = g$, and **(d)** if $g \in G$, then there is an element $g^{-1} \in G$ such that $g \cdot g^{-1} = g^{-1} \cdot g = e$.

7.6   Properties of Relatively Prime Integers

It is a fundamental fact of elementary number theory that if a, b are relatively prime, then we can find other integers x and y such that

$$xa + yb = 1. \tag{$*$}$$

For example, we have noted that $a = 72$ and $b = 175$ are relatively prime. The corresponding integers x, y are $x = -17$ and $y = 7$. Thus

$$(-17) \cdot 72 + 7 \cdot 175 = 1.$$

This result follows from our Theorem 7.2.5. Here we give another proof. In fact we use Fermat's theorem above. For, since b is relatively prime to a, we see that

$$b^{\varphi(a)} = 1 \bmod a.$$

But this just says that

$$b^{\varphi(a)} - 1 = k \cdot a$$

for some integer k. Unraveling this equation gives $(*)$.

In practice, one finds x and y using the Division Algorithm (otherwise known as long division).

In the example of 72, 175, one calculates:

$$
\begin{aligned}
175 &= 2 \cdot 72 + 31 \\
72 &= 2 \cdot 31 + 10 \\
31 &= 3 \cdot 10 + 1.
\end{aligned}
$$

You know you are finished when the remainder is 1.

For now we have

$$
\begin{aligned}
1 &= 31 - 3 \cdot 10 \\
&= 31 - 3 \cdot (72 - 2 \cdot 31) \\
&= 7 \cdot 31 - 3 \cdot 72 \\
&= 7 \cdot (175 - 2 \cdot 72) - 3 \cdot 72 \\
&= 7 \cdot 175 - 17 \cdot 72.
\end{aligned}
$$

That is the decomposition we seek.

EXERCISES

1. Calculate gcd(455, 1235).

2. Use the sieve of Eratosthenes to make a list of all primes up to 100.

3. Prove that there are infinitely many primes of the form $6x - 1$.

* **4.** Let $\psi(x)$ be the number of primes of the form $4x - 1$ that are $\leq x$. Use a computer to make a conjectural guess about the limit as $x \to \infty$ of $\psi(x)/\pi(x)$.

5. Let a, b, c, n be integers. Prove that

   (a) If $a|n$ and $b|n$ with $\gcd(a, b) = 1$, then $ab|n$.
   (b) If $a|bc$ and $\gcd(a, b) = 1$, then $a|c$.

6. Let a, b, c, d, m be integers. Prove that

   (a) If $a|b$ and $b|c$, then $a|c$.
   (b) If $a|b$ and $c|d$, then $ac|bd$.
   (c) If $m \neq 0$, then $a|b$ if and only if $ma|mb$.
   (d) If $d|a$ and $a \neq 0$, then $|d| \leq |a|$.

7. For each of the following, apply the Division Algorithm to find q and r such that $a = bq + r$ and $0 \leq r < |b|$.

   (a) $a = 300, b = 17$
   (b) $a = 729, b = 31$
   (c) $a = 300, b = -17$
   (d) $a = 389, b = 4$

8. Suppose that a, b, n are positive integers. Prove that if $a^n|b^n$, then $a|b$.

9. Prove that, if a positive integer n is a perfect square, then n cannot be written in the form $4k + 3$ for k an integer. [**Hint:** Compute the remainder upon division by 4 of each of $(4m)^2$, $(4m+1)^2$, $(4m+2)^2$, $(4m + 3)^2$.]

10. Prove that no integer in the sequence

$$11 \,,\, 111 \,,\, 1111 \,,\, 11111 \,,\, 111111 \,,\, \ldots$$

is a perfect square. [**Hint:** $111 \cdots 111 = 111 \cdots 108 + 3 = 4k + 3$.]

11. Prove that, for any positive integer n, the set $[\mathbb{Z}/n\mathbb{Z}]^*$ (the integers modulo n which are relatively prime to n) with the binary operation of multiplication is a group.

12. Compute the following gcds.

   (a) $\gcd(15, 5)$

   (b) gcd(247, 299)

   (c) gcd(51, 897)

   (d) gcd(136, 304)

13. Find $x, y \in \mathbb{Z}$ such that $121x + 55y = 11$.

14. Prove that, if a and b are integers and p is a prime, then $(a + b)^p = a^p + b^p \mod p$. [**Hint:** Of course you may assume that the binomial coefficient $\binom{n}{k}$ is an integer.]

15. Prove that if x, y is a solution of the equation $ax + by = d$ with $d = \gcd(a, b)$, then, for all $c \in \mathbb{Z}$,

$$x' = x + c \cdot \frac{b}{d} \quad, \quad y' = y - c \cdot \frac{a}{d} \qquad (*)$$

   is also a solution of $ax + by = d$.

16. Refer to Exercise 15. Find the solution to $51x + 119y = 17$.

17. Prove that a number $n \in \mathbb{Z}$ is divisible by 3 if and only if the sum of the digits of n is divisible by 3.

* 18. With reference to the Chinese Remainder Theorem, find a positive integer n that satisfies these congruences:

$$
\begin{array}{rcl}
n = 3 & \mathrm{mod} & 17 \\
n = 2 & \mathrm{mod} & 23 \\
n = 5 & \mathrm{mod} & 11 \\
n = 19 & \mathrm{mod} & 37
\end{array}
$$

* 19. Let $f(x) = x^2 + ax + b$ be a quadratic polynomial with integer coefficients. For example, f could be $f(x) = x^2 - 7x + 5$. Formulate a conjecture about when the set

$$S = \{f(n) : n \in \mathbb{Z} \text{ and } f(n) \text{ is prime}\}$$

   is infinite. Give numerical evidence (perhaps using a computer) to support your conjecture.

20. Find an integer n so that $37n = 1 \mod 101$.

21. Let p be a prime. Prove that $\mathbb{Z}/p\mathbb{Z}$ (the integers modulo p) is a field.

22. Show that, if n is a positive integer so that n and $n^2 + 2$ are prime then $n = 3$.

23. Let S be the set $\mathbb{Z}/5\mathbb{Z}$ (the integers modulo 5) with binary operation multiplication. Is this a group?

24. Find all four solutions of the equation

$$x^2 - 1 = 0 \mod 35.$$

25. Compute the last digit of 3^{45}.

* **26.** For $n \in \mathbb{N}$, let $\sigma(n)$ denote the sum of the divisors of n. For example, $\sigma(6) = 1 + 2 + 3 + 6 = 12$ and $\sigma(10) = 1 + 2 + 5 + 10 = 18$. Assume that $n = pqr$, with p, q, r distinct primes. Devise an algorithm that uses n, $\varphi(n)$, and $\sigma(n)$ to determine the prime factorization of n.

As an example of what we are asking, let $n = 105$. Then $p = 3$, $q = 5$, $r = 7$. So the input to the requested algorithm would be

$$n = 105 \quad , \quad \varphi(n) = 48 \quad , \quad \sigma(n) = 192 .$$

And the output would be $3, 5, 7$.

27. Given a example to show that it is *not* true that

$$(a \cdot b) \bmod n = (a \bmod n) \cdot (b \bmod n) .$$

In fact you should be able to give an example with $n = 3$.

8

Zero-Knowledge Proofs

8.1 Basics and Background

Modern security considerations make it desirable for us to have new types of encryption schemes. It is no longer enough to render a message so that only the intended recipient can read it (and outsiders cannot). In today's complex world, and with the advent of high-speed digital computers, there are new demands on the technology of cryptography. The present chapter will discuss some of these considerations.

In the old days (beginning even with Julius Caesar), it was enough to have a method for disguising the message that we were sending. For example, imagine that the alphabet is turned into numeric symbols by way of the scheme

$$
\begin{aligned}
A &\longmapsto 0 \\
B &\longmapsto 1 \\
C &\longmapsto 2 \\
&\vdots \\
X &\longmapsto 23 \\
Y &\longmapsto 24 \\
Z &\longmapsto 25.
\end{aligned}
$$

Then use the encryption

$$ n \longmapsto n + 3 \bmod 26. \qquad (\star) $$

And now convert these numbers back to roman letters. As a simple example, the phrase

WHAT ME WORRY?

translates to the string of integers

22 7 0 19 12 4 22 14 17 17 24

DOI: 10.1201/9781003214564-8

Notice that it is common, in elementary cryptography, to ignore punctuation and spaces.

The encryption ($\star$) turns this string of integers into

$$25 \quad 10 \quad 3 \quad 22 \quad 15 \quad 7 \quad 25 \quad 17 \quad 20 \quad 20 \quad 2$$

and this, in turn, transliterates to

$$\text{ZKDWPHZRUUC} \hspace{4cm} (\star\star)$$

It is clear that anyone receiving the message ($\star\star$) would have no idea what it means—nor even what the message was about, or what context it fits into. On the other hand, such an encoded message is pretty easy to decrypt. Especially if the decrypter knows that we have used a simple "shift algorithm" to encode the message, and if in addition he/she knows that the most commonly used letters in the alphabet are E and then T, then it would be a fairly simple matter to reverse-engineer this encryption and recover the original message.

Today life is more complex. One can imagine that there would be scenarios in which

(1) You wish to have a means that a minimum-wage security guard (whom you don't necessarily trust) can check that people entering a facility know a password—but you don't want him to know the password.

(2) You wish to have a technology that allows anyone to encrypt a message—using a standard, published methodology—but only someone with special additional information can decrypt it.

(3) You wish to have a method to be able to convince someone else that you can perform a procedure, or solve a problem, or prove a theorem, without actually revealing the details of the process.

This may all sound rather dreamy, but in fact—thanks to the efforts and ideas of C. Cocks, R. Merkle, W. Diffie, and M. Hellman, followed by R. Rivest, A. Shamir, and L. Adleman—it is now possible. The idea of public key encryption is due to the first group. The contribution of the second group was to make the methodology more secure.

The so-called RSA encryption scheme is now widely used. For example, the e-mail messages that I receive on my cell phone are encrypted using RSA. Banks, secure industrial sites, high-tech government agences (for example, the National Security Agency), and many other parts of our society routinely use RSA to send messages securely.

In this discussion, we shall describe how RSA encryption works, and we shall encrypt a message using the methodology. We shall describe all the mathematics behind RSA encryption, and shall prove the results necessary to flesh out the theory behind RSA. We shall also describe how to convince someone that you can prove the Riemann hypothesis—without revealing any details of the proof. This is a fascinating idea—something like convincing your mother that you have cleaned your room without letting her have a look at the room. This idea has profound and far-reaching applications.

It should be noted that the effectiveness of the RSA encryption system hinges on the fact that calculating the prime factorization of a very large integer (300 digits or more) is computationally expensive. It takes quite a long time, even on a fast computer. When quantum computing has been achieved, and this could happen in our lifetimes, then this will no longer be an effective deterrent. There are newer lattice-based encryption systems that will avoid this problem.

8.2 Preparation for RSA

8.2.1 Background ideas

We now sketch the background ideas for RSA. These are all elementary ideas from basic mathematics. We shall also use some elementary number theory from the last chapter. It is remarkable that these are all that are needed to make this profound new idea work.

8.2.2 Computational complexity

• Suppose that you have a deck of N playing cards and you toss them in the air. Now you want to put them back into their standard order. How many "steps" will this take? [We want to answer this question in such a manner that a machine could follow the instructions.]

First we look through all N cards until we find the first card in the ordering.

Then we look through the remaining $N - 1$ cards until we find the second card in the ordering.

And so forth.

So the reordering of the cards takes at most

$$N + (N - 1) + (N - 2) + \cdots 3 + 2 + 1 = \frac{N(N + 1)}{2}$$

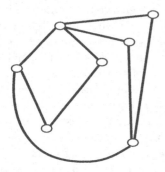
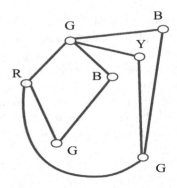

A typical planar graph.   An admissible coloring for
   the planar graph.

FIGURE 8.1
Coloring a graph.

steps. Notice that this answer is a quadratic polynomial in N. Thus we say that the problem can be solved in polynomial time.

• We have heard a rumor that the four-color theorem is true. So we have a graph with N vertices and we wish to color each vertex, using either red, yellow, blue, or green. See Figure 8.1, where we have used R, G, B, Y to denote the colors red, green, blue, and yellow. The only rule is that two adjacent vertices (i.e., vertices that are connected by a segment) cannot be the same color.

Of course the number of possible colorings is the number of functions from the set with N objects to the set with 4 objects. That is 4^N. The machine, being as dumb as it is, will simply try all the possible colorings until it finds one that works. Thus we see that the number of steps will be an exponential function of N.

We call this an exponential time problem.

• Another interesting exponential time problem is that of scheduling planes for an airline. If you have n cities and k planes and you take into account different populations, different demands, crew availability, fuel availability, and other factors, then it is easy to convince yourself that this is a problem of exponential complexity. The theory of linear programming (developed by L. Kantorovich, T. C. Koopmans, and F. L. Hitchcock and later turned into an entire field by G. B. Dantzig) can be used to reduce many problems of this kind to polynomial complexity. Linear programming is routinely used by the airlines for this purpose. See [DAM1], [DAM2].

• Certainly one of the most famous exponential time problems is the "traveling salesman problem." The issue here is that a certain salesman wants to visit each of n cities precisely once. What is his most efficient path? It is not difficult to discern that there are exponentially many possible routes, and no evident strategy for picking one in any efficient manner.

8.3 The RSA System Enunciated

8.3.1 Description of RSA

Now we can quickly and efficiently describe how to implement the RSA encryption system, and we can explain how it works.

Imagine that the President has an important message that she wishes to send to the Secretary of Defense. Of course the Secretary of Defense is a highly placed person of many responsibilities, and you can imagine that the President's message is quite secret. So she wants to encode the message:

Your time is up. Hasta la vista, baby.

So the President goes to the library and finds the RSA encryption book. This is a readily available book that anyone can access. It is not secret. A typical page in the book reads like this:

NAME	VALUE OF n	VALUE OF e
Puck, Wolfgang	4431 ... 7765	8894 ... 4453
Rehnquist, William	6668 ... 2345	1234 ... 9876
Riddle, Nelson	7586 ... 2390	4637 ... 4389
Rin Tin-Tin	5355 ... 5353	5465 ... 7648
Rogers, Roy	7859 ... 4359	3058 ... 2934
Roosevelt, Theodore	7835 ... 2523	7893 ... 4232
Rotten, Johnny	3955 ... 4343	4488 ... 9922
Roy, Rob	3796 ... 5441	2219 ... 3319
Russert, Tim	6464 ... 4646	3223 ... 3232
Schwarzenegger, Arnold	6894 ... 3242	7525 ... 2314
Secretary of Defense	1117 ... 8854	9266 ... 2388
Simpson, Orenthal James	6678 ... 2234	4856 ... 2223

What does this information mean? Of course we know, thanks to Euclid, that there are infinitely many primes. So we can find prime numbers with as many digits as we wish. Each number n in the RSA encryption book is the product of two 150-digit primes p and q: Thus $n = p \cdot q$. Each number e is chosen to be a number with at least 200 digits that is relatively prime to $\varphi(n) = (p-1) \cdot (q-1)$. Of course we do not publish the prime factorization of the number n; we also do not publish $\varphi(n)$. All that we publish is n and e for each individual.

Now an important point to understand is that the President does not need to understand any mathematics or any of the theory of RSA encryption in order to encode her message. [Well, it would be nice if she understood modular arithmetic. But she is, after all, the President of the United States.] All she does is this:

(1) First she breaks the message into units of five letters. We call these "words," even though they may not be English language words.

For the message from the President to the Secretary of Defense, the "words" would be

YOURT   IMEIS   UPHAS   TALAV   ISTAB
ABY

(2) She transliterates each "word" into a sequence of numerical digits, using our usual scheme of translation ($A \to 0$, $B \to 1$, $C \to 2$, etc.).

(3) Then she encodes each transliterated word w with the rule

$$w \longmapsto w^e \bmod n \equiv \widetilde{w}.$$

The President will send to the Secretary of Defense this sequence of encrypted words $\widetilde{w}$. That is all there is to it.

The real question now is:

What does it take to decrypt the encoded message? How can the Secretary of Defense read the message?

This is where some mathematics comes into the picture. We must use the Euler–Fermat theorem, and we must use our ideas about relatively prime integers. But the short answer to the question is this. If $\widetilde{w}$ is a word encrypted according to the simple scheme described above, then we decrypt it with this algorithm:

We find integers x and y so that $xe + y\varphi(n) = 1$ and then we calculate

$$\widetilde{w}^x \bmod n.$$

That will give the decrypted word w with which we began. [We shall provide the mathematical details of this assertion in the next section.] Since w has only five characters, and n has 300 characters, we know that $w \bmod n = w$—so there is no ambiguity arising from modular arithmetic. We can translate w back into roman characters, and we recover our message.

Now here is the most important point in our development thus far:

In order to **encrypt** a message, we need only look up n and e in the public record RSA encryption book. But, in order to decrypt the message, we must know x. Calculating x necessitates knowing $\varphi(n)$, and that necessitates knowing the prime factorization of n.

It is a theorem that calculating the prime factorization of an integer with k digits is a problem of exponential complexity in k. For an integer with 300 digits, using a reasonably fast computer, it would take perhaps a year to find the prime factorization.[1]

8.4　The RSA Encryption System Explicated

8.4.1　Explanation of RSA

In fact, with all the preliminary setup that we have in place, it is a simple matter to explain the RSA encryption system.

For suppose that we selected an $n = p \cdot q$ and an e relatively prime to $\varphi(n) = (p-1) \cdot (q-1)$ corresponding to a particular person listed in the RSA encryption book. If we are the certified decrypter, then we know the prime factorization of n—that is, we know that $n = p \cdot q$ for p and q prime.

We therefore know $\varphi(n) = p \cdot q - p - q + 1 = (p-1) \cdot (q-1)$, and so we can calculate the x and the y in the identity $xe + y\varphi(n) = 1$. Once we know x, then we know everything. For

---

[1] A bright student who heard my lecture on this topic pointed out that—if your message is just five letters long—then there are only 26^5 possible encryptions. You could decrypt the message by trial and error. Because of considerations like this, we often find it convenient to append a 50-digit random integer to the message. This technique is discussed in detail near the end of the chapter.

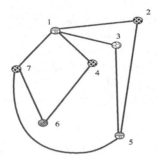

FIGURE 8.2
A colored graph.

$$
\begin{aligned}
\widetilde{w}^x \bmod n &= \left[w^e\right]^x \bmod n \\
&= w^{ex} \bmod n \\
&= w^{1-y\varphi(n)} \bmod n \\
&= w \cdot \left[w^{\varphi(n)}\right]^{-y} \bmod n \\
&= w \cdot 1^{-y} \bmod n \\
&= w \bmod n \\
&= w
\end{aligned}
$$

since w is certainly relatively prime to n.

This shows how we recover the original word w from the encrypted word $\widetilde{w} = w^e \bmod n$.

See https://www.britannica.com/topic/RSA-encryption for more about RSA encryption.

8.5   Zero-Knowledge Proofs

8.5.1   How to keep a secret

We shall now give a quick and dirty description of how to convince someone that you can prove **Proposition A** without revealing any details of the proof of **Proposition A**. The idea comes across most clearly if we deal again with coloring of graphs.

So suppose once again that we are given a graph. See Figure 8.2. We are the prover, and there is a remotely located verifier.

It is our job to convince the verifier that we know how to 4-color this graph. But we do not want the verifier to actually *know* how to color the graph. We only want him to be *convinced* that *we know how to do it.*

We begin by transmitting the adjacency matrix to the verifier. You can see that the adjacency matrix has a 1 in position (j, k) if the jth vertex and the kth vertex are connected. It has a 0 in position (j, k) if the jth vertex and the kth vertex are disconnected. This data is exhibited in Table 8.3.

This transmission is straightforward, and need not be encoded. We simply tell the verifier: "In position $(1, 1)$ of the matrix there is a 1"; "In position $(1, 2)$ of the matrix there is a 0"; In position $(1, 5)$ of the matrix there is neither." And so forth. That way the verifier can draw a picture of the graph.

Now we number the colors 1, 2, 3, 4. Here

$$1 \leftrightarrow \text{blue}$$
$$2 \leftrightarrow \text{red}$$
$$3 \leftrightarrow \text{yellow}$$
$$4 \leftrightarrow \text{green}$$

TABLE 8.3

	1	2	3	4	5	6	7
1	1	1	1	1	0	0	1
2	1	1	0	0	1	0	0
3	1	0	1	0	1	0	0
4	1	0	0	1	0	1	0
5	0	1	1	0	1	0	1
6	0	0	0	1	0	1	1
7	1	0	0	0	1	1	1

Finally, imagine that our coloring of the graph is as in Figure 8.2. Of course we have used shading to suggest the coloring. We wish to communicate to the verifier that we have a valid coloring—in such a manner that she can check it, but she cannot learn any of the details of the coloring.

As we see from Figure 8.2, the coloring is encoded as

$$12 \quad 23 \quad 34 \quad 43 \quad 52 \quad 61 \quad 73 \,.$$

This is read as "node 1 is colored with color 2 (red)," "node 2 is colored with color 3 (yellow), etc. We will transmit these pairs of digits, suitably encoded, to the verifier.

The trouble is that the verifier already knows that there are only seven nodes in the graph, and only four colors, so the verifier could (with a little effort) figure out what color has been assigned to what node just with a little trial and error—even though the information has been encoded. So this will not do.

Thus, instead of encoding and sending 12 and 23 and 34 and so forth, instead the prover encodes and sends

$$12\, r_1$$
$$23\, r_2$$
$$34\, r_3$$
$$\textbf{etc.,}$$

where r_1, r_2, r_3 are 50-digit random integers.

More precisely, the step-by-step scenario is this:

(1) The prover sends the entire coloring to the verifier, in encoded form as indicated above.

(2) The verifier stares at the adjacency matrix. The verifier notices that, for example, vertices 4 and 6 are adjacent. And then inquires specifically about those two vertices.

(3) The prover sends the verifier the coloring for those two particular vertices (with a 50-digit random integer attached, just as before).

(4) The verifier encrypts the information for those two vertices—using the pre-agreed-upon public key encryption system. The verifier then checks that those two vertex colorings match colorings in the full coloring of the graph that was sent in **Step (1)**.

Now the verifier has checked that one pair of adjacent vertices is suitably colored (i.e., with different colors). If the verifier wants to perform further verifications, then the preceding steps are repeated. Except that first the prover assigns numbers to the four colors **red, yellow, blue,** and **green** in some new random fashion. And the prover chooses an entirely new set of random 50-digit integers. Then the prover sends the entire colored graph, gets a query from the verifier, and so forth.

If there are n nodes to be colored, then there are $n(n-1)/2$ possible pairs of nodes. The probability that the prover *lied* about the coloring and that the verifier—in asking for the coloring of a particular pair of nodes—failed to catch the lie is $1/[n(n-1)]$. If the entire process is iterated again, then the probability that the verifier failed to catch the lie is $1/[n^2(n-1)^2]$. And so forth. Thus each successive verification increases the likelihood that the verifier may be certain of his check.

The point of this procedure is that the verifier can check that any pair of adjacent vertices is colored correctly, that no two adjacent vertices are colored the same, but the verifier cannot amalgamate the information and produce the entire coloring of the graph.

Of course the example we have presented is for graph coloring, just because that is simple to describe. But any proof whatsoever can be translated into binary code and then rendered as a statement about the coloring of some graph. So, in fact. the example we have given is perfectly general.

8.6 Concluding Remarks

The RSA encryption scheme is one of the great ideas of modern coding theory. It is being developed and enhanced even now. There are versions for multiple verifiers, for dishonest provers, and many other variants. The history of RSA is a remarkable one. There was a talk at the 1986 International Congress of Mathematicians in Berkeley about the method. After that, the government attempted to co-opt the method, retract all the preprints, and suppress the information. Interestingly, it was the National Security Agency (the branch of the government in Washington that specializes in cryptography) that stepped in and prevented the government intervention. Now RSA is in the public domain, and anyone can use it. It is a powerful tool.

Exercises

1. Perform an RSA encryption of the message

   The Beatles forever.

   using $n = 34161 = 177 \times 193$ and $e = 5^2 \times 7^2$.

2. Use the RSA method to *decrypt* the encoded message that you created in Exercise 8.1. Verify that the decrypted message is "The Beatles forever."

3. Devise a method for convincing a remote verifier that you can decrypt a certain message, without revealing to him/her what the decrypted message is.

4. Prove that the RSA encryption of the product of two messages is equal to the product of the encryptions of the individual messages. This fact can sometimes be used to speed up the breaking of RSA code.

5. Is there any value in performing RSA encryption twice? That is to say, take a message, RSA-encrypt it, and then RSA-encrypt that encrypted message.

6. The way we have described RSA encryption, it is entirely possible for someone to pretend to be you, encrypt a message, and send

it to your collaborator. They would simply use the same encoding information (*e* and *n*) that you use. What device could you use so that someone receiving an RSA-encrypted message that purports to be from you *really is* from you?

7. The ciphertext

   QBB JXU MEHBT YI Q IJQWU QDT QBB JXU CUD QDT
   MECUD CUHUBO FBQOUHI

   has been generated by an advanced Caesar cipher with shift 16 (i.e., "A" is mapped to "Q", etc.). Decrypt it.

* 8. The Greek Polybius (∼200–118 B.C.E.) invented a monoalphabetic cipher that converts alphabetical characters into numerical characters. It is based on a checkerboard:

#	1	2	3	4	5
1	a	b	c	d	e
2	f	g	h	ij	k
3	l	m	n	o	p
4	q	r	s	t	u
5	v	w	x	y	z

We associate to each letter of the alphabet a two-digit number by looking at the row and column that it appears in. For example, *k* is associated to 25 and *t* is associated to 44.

Use the Polybius checkerboard to encrypt the message.

   God is in the details.

9. The Playfair Cipher replaces single letters with digraphs in an interesting manner. We begin with a keyword, say TURKEY. Then we create a 5 × 5 matrix by beginning with the key word and then filling in all the other letters in order (and we combine I and J). The result is

T	U	R	K	E
Y	A	B	C	D
F	G	H	I/J	L
M	N	O	P	Q
S	V	W	X	Z

The message is now broken up into pairs of letters (called *digraphs*), where an *X* is inserted between identical letters and also at the end of the message if necessary (to make an even number of letters).

Suppose that the message is

THE DAY IS SUNNY.

Then the digraph decomposition is

TH ED AY IS XS UN NY

Now all digraphs fall into one of three categories: **(i)** both letters in the same row, **(ii)** both letters in the same column, **(iii)** neither. We treat these as follows:

(a) If both letters are in the same row, then each is replaced by the letter to its right in the matrix. But if one of the letters occurs in the last position of the row, then it is replaced by the the letter at the beginning of the row. So, for example, BR becomes IU.

(b) If both letters are in the same column, then each is replaced by the letter beneath it in the matrix. But if one of the letters is at the bottom of the column, then it is replaced by the letter at the top of the column. So, for instance, RS becomes YL.

(c) If the letters in a digraph are neither in the same row nor in the same column, then we follow these rules:

  (i) To encipher the first letter, look along its row until you reach the column containing the second letter. The letter at the intersection is the one that you use as the replacement.

  (ii) To encipher the second letter, look along its row until you find the column containing the first letter; the letter at this intersection then is the replacement for the second letter.

Use this technique to encrypt the message enunciated above.

10. Does RSA encryption work if p and q are not prime?

11. Use the Division Algorithm to find the greatest common divisor of 248 and 164.

12. Use the Division Algorithm to find integers x and y so that

$$27x + 64y = 1.$$

13. Suppose you are person A, and you have chosen as your two primes $p = 97$ and $q = 173$, and you have chosen $e = 5$. Thus you told B that $n = 16781$ (which is just pq) and you told him that $e = 5$. He encodes a message (a number) for you and tells you that the encoding is 5347. Can you figure out the original message?

14. We know from the text that Fermat's Little Theorem says this:

  Theorem: If p is a prime integer and $1 \le a < p$ is another integer, then

$$a^{p-1} \equiv 1 \bmod p. \qquad (*)$$

If we are given a positive integer p, and we want to test whether p is prime, we can just pick an integer a from the interval $[1, p-1]$ at random and see whether $(*)$ holds. If the equality fails for some particular value of a, then we may definitely conclude that p is composite. If the equality holds for some particular value of a, then we can say that p is "probably" prime.

Test the integer 13 for primality, using $a = 4$ and $a = 5$.

15. Refer to Exercise 8.14 for terminology. Apply the Fermat test to the number 14 to test for primality. Use three different values for a.

16. The Chinese Remainder Theorem (see Exercises 8.19 and 8.20) says that there is an integer with residue 1 modulo 2, residue 2 modulo 3, and residue 3 modulo 5. Find it.

17. If p is a prime and q is a distinct prime, them $p^2 - q^2$ will never be prime. But $p^2 + q^2$ could be prime. Explain.

18. Do a `Google` search to learn what are the most commonly used letters in the roman alphabet in the English language. What are the most commonly used digraphs in the roman alphabet in the English language? How can one use this information in cryptography?

19. The most classical version of the Chinese Remainder Theorem says that, if $p_1, \ldots, p_k$ are distinct primes and $a_1, a_2, \ldots, a_k$ are positive integers, then there exists a solution x to the simultaneous equations

$$x = a_j \bmod p_j, \quad j = 1, \ldots, k.$$

Find a constructive proof of this result.

20. A version of the Chinese Remainder Theorem says this:

Let p and q be two numbers (not necessarily primes), but which are relatively prime. Then, if $a = b \bmod p$ and $a = b \bmod q$, we may conclude that $a = b \bmod pq$.

Prove this theorem.

9

Examples of Axiomatic Theories

9.1 Group Theory

We begin this chapter with a brief discussion of an axiomatic structure that may be new to you. After set theory, it is one of the most pervasive in all of mathematics.

Definition 9.1.1 Let G be a set, and let $P : G \times G \to G$ be a function. We call the ordered pair (G, P) a *group* if the following axioms are satisfied (following custom, we shall usually write $P(g, h)$ as $g \cdot h$):

1. Associativity If $g, h, k \in G$, then $g \cdot (h \cdot k) = (g \cdot h) \cdot k$;

2. Identity Element There is a distinguished identity element $e \in G$ such that, for all $g \in G$, $e \cdot g = g \cdot e = g$.

3. Multiplicative Inverse For each $g \in G$ there is an element $h \in G$ such that $g \cdot h = h \cdot g = e$.

It is common to denote the inverse element specified in Axiom **3** by g^{-1}, and we shall do so in what follows.

Notice that we do *not* assume that a group is commutative; that is, we do not assume that $g \cdot h = h \cdot g$ for all $g, h \in G$. The property of associativity that we postulate in Axiom 1 is a different property: it says that when we are combining three elements we may group them, two by two, in either of the two obvious ways; the same answer results.

EXAMPLE 9.1.2 Let G be the positive real numbers, and let the group operation be multiplication: $P(x, y) = x \cdot y$, where $\cdot$ is ordinary multiplication of reals. Then (G, P) is a group.

Axiom 1: Of course multiplication of real numbers is associative.
Axiom 2: The number 1 is the identity element for multiplication: $1 \cdot x = x \cdot 1 = x$ for any real number x.
Axiom 3: The multiplicative inverse of a group element is its ordinary reciprocal. That is, if $x \in \mathbb{R}$ satisfies $x > 0$, then $1/x$ is its multiplicative inverse.
◇

DOI: 10.1201/9781003214564-9

EXAMPLE 9.1.3 Let G be the integers, and let $P(x,y) = x + y$ (ordinary addition). Then (G, P) is a group.

Axiom 1: Certainly addition of integers is associative.
Axiom 2: The number 0 is the additive identity.
Axiom 3: The additive inverse of a group element is its negative: if $m \in \mathbb{Z}$, then $-m$ is its group inverse.   ◇

EXAMPLE 9.1.4 Let G be the $k \times k$ matrices with real entries and non-zero determinant. This is sometimes called the *general linear group on k letters* and is denoted by $GL(k, \mathbb{R})$.

  Let P be ordinary matrix multiplication. Then (G, P) is a group.

Axiom 1: Matrix multiplication is associative.
Axiom 2: The group identity is the matrix

$$
I_k \equiv \left. \begin{pmatrix} 1 & 0 & 0 & \dots & 0 \\ 0 & 1 & 0 & \dots & 0 \\ & & \dots & & \\ 0 & 0 & \dots & 1 & 0 \\ 0 & 0 & \dots & 0 & 1 \end{pmatrix} \right\} k
$$
$$
\underbrace{\phantom{\begin{pmatrix} 1 & 0 & 0 & \dots & 0 \end{pmatrix}}}_{k}
$$

Thus, if $m \in G$, then $I_k \cdot m = m \cdot I_k = m$.
Axiom 3: The multiplicative inverse of a group element is its matrix inverse. Thus, if $m \in G$, then the inverse matrix m^{-1} is the group inverse.

  Notice in this example that it is important to restrict attention to square matrices, so that multiplication of any two elements in any order will make sense. We also require that each matrix have non-zero determinant, so that each matrix will have an inverse. To see that G is closed under the group operation of matrix multiplication, we must note that if $M, N \in G$, then $\det(M \cdot N) = (\det M)(\det N) \neq 0$.   ◇

  Unlike the previous two examples, this last one is a noncommutative group. Commutative groups are usually referred to as *abelian*, in honor of N. H. Abel (1802–1829), who first studied them in detail. The book [HAL] gives a solid introduction to classical group theory.

  The advantage of the axiomatic method, in the present context, is that when we prove a proposition or theorem about "a group G," it applies simultaneously to all groups. Thus the axiomatic method gives us both a way of being concise and a way of cutting to the heart of the matter.

Proposition 9.1.5 *The multiplicative identity for a group is unique.*

Proof: Let G be a group. Let e and e' both be elements of G that satisfy Axiom 2. Then

$$
e = e \cdot e' = e'.
$$

Thus e and e' must be the same group element. $\square$

Proposition 9.1.6 *Let G be a group and $g \in G$. Then there is only one multiplicative inverse for g.*

Proof: Suppose that h and k both satisfy the properties of the multiplicative inverse (Axiom 3) relative to g. Then

$$h = h \cdot e = h \cdot (g \cdot k) = (h \cdot g) \cdot k = e \cdot k = k.$$

Thus h and k must be the same group element, establishing that the multiplicative inverse is unique. $\square$

Niels Henrik Abel (1802–1829)

In 1815, Niels Abel and his older brother were sent to the Cathedral School in Christiania. When a new mathematics teacher Bernt Holmboë joined the school in 1817, things changed markedly for Abel. Abel began to study university level mathematics texts and, within a year of Holmboë's arrival, Abel was reading the works of Euler, Newton, Lalande and d'Alembert.

Abel was able to enter the University of Christiania in 1821. He graduated in 1822. While in his final year at school, however, Abel had begun working on the solution of quintic equations by radicals. In 1823 Abel published papers on functional equations and integrals. In Abel's third paper, he gave the first solution of an integral equation.

Abel began again to work on quintic equations and, in 1824, he proved the impossibility of solving the general equation of the fifth degree in radicals. He published the work in French and at his own expense since he wanted an impressive piece of work to take with him when he was on his travels.

In August 1825, Abel was given a scholarship from the Norwegian government to allow him to travel abroad. Abel met Crelle in Berlin and the two became firm friends. This proved the most useful part of Abel's whole trip, particularly as Crelle was about to start up a new journal. Abel was encouraged by Crelle to write a clearer version of his work which was published in 1827 in the first volume of Crelle's Journal, along with six other papers by Abel.

Abel had just finished an extensive treatise on a certain class of transcendental functions. He showed it to Cauchy, but the man scarcely deigned to glance at it. Abel's theorem is a vast generalization of Euler's relation for elliptic integrals. Two referees, Cauchy and Legendre, were appointed to referee the paper and Abel remained in Paris for a few months emaciated, gloomy, weary, and constantly worried. With his health in a very poor state, he returned to Berlin at the end of 1826.

In Berlin, Abel borrowed some money and continued working on elliptic functions. He wrote a paper in which he radically transformed the theory of elliptic integrals to the theory of elliptic functions by using their inverse functions. Crelle tried to persuade Abel to remain in Berlin until he could find an academic post for him and he even offered Abel the editorship of Crelle's Journal. He returned to Christiania in May 1827 and was awarded a small amount of money by the university.

In 1828, Abel was shown a paper by Jacobi on transformations of elliptic integrals. Abel quickly showed that Jacobi's results were consequences of his own and added a note to this effect to the second part of his major work on elliptic functions. He put this to one side to compete with Jacobi in the theory of elliptic functions, quickly writing several papers on the topic.

Abel travelled by sled to visit his fiancée again in Froland for Christmas 1828. He became seriously ill on the sled journey. Crelle was told and he redoubled his efforts to obtain an appointment for Abel in Berlin. He succeeded and wrote to Abel on the 8 April 1829 to tell him the good news. It was too late; Abel had already died.

Proposition 9.1.7 *Let g be an element of the group G. Then*

$$\left(g^{-1}\right)^{-1} = g.$$

Proof: Observe that

$$g \cdot g^{-1} = e$$

and

$$g^{-1} \cdot g = e.$$

Thus g satisfies the properties of the inverse element (Axiom 3) relative to g^{-1}. Since the last proposition establishes that the inverse element for g is unique, it follows that g must be the multiplicative inverse for g^{-1}. In other words, $\left(g^{-1}\right)^{-1} = g$. □

Proposition 9.1.8 *Let g, h be elements of a group G. Then $(g \cdot h)^{-1} = h^{-1} \cdot g^{-1}$.*

Proof: We calculate that

$$
\begin{aligned}
\left[h^{-1} \cdot g^{-1}\right] \cdot [g \cdot h] &= h^{-1} \cdot \left[g^{-1} \cdot [g \cdot h]\right] \\
&= h^{-1} \cdot \left[(g^{-1} \cdot g) \cdot h\right] \\
&= h^{-1} \cdot \left[e \cdot h\right] \\
&= h^{-1} \cdot h \\
&= e.
\end{aligned}
$$

A similar calculation shows that

$$[g \cdot h] \cdot \left[h^{-1} \cdot g^{-1}\right] = e.$$

The assertion follows. □

Definition 9.1.9 Let G be a group and $H \subset G$. We call H a *subgroup* of G if the following three properties hold:

(1)   Closure The group operation P associated with G satisfies $P : H \times H \to H$. In other words, H is closed under the group operation of G (see Section 5.3 for the concept of "closed");

(2)   Identity Element The group identity e is an element of H;

(3)   Multiplicative Inverse If $h \in H$, then its group inverse element h^{-1} lies in H.

Notice that the point of the last definition is that H is itself a group, using operations (and the group identity) inherited from the larger group G.

EXAMPLE 9.1.10 The pair $(\mathbb{Q}, +)$ of the rational numbers under the ordinary operation of addition forms a group. Then the integers $\mathbb{Z} \subset \mathbb{Q}$ form a subgroup. That is, the integers are a group under the same operation of addition. They are closed under this operation. $\quad\diamond$

EXAMPLE 9.1.11 Let G be the 3×3 matrices with real entries and non-vanishing determinant.

$$\begin{pmatrix} a & b & c \\ d & e & f \\ g & h & i \end{pmatrix}.$$

Let the group law be matrix multiplication. As we noted in Example 9.1.4, this is a group. Let H be the subset of G consisting of those matrices with non-zero entries on the main diagonal and zero entries off the diagonal:

$$\begin{pmatrix} a & 0 & 0 \\ 0 & e & 0 \\ 0 & 0 & g \end{pmatrix}.$$

Then H is a subgroup. $\quad\diamond$

Let G be a group and H a subgroup. Let us define a relation $\mathcal{R}$ on G as follows: $(x, y) \in \mathcal{R}$ provided that $x^{-1} \cdot y \in H$. If x is any element of G, then of course $x^{-1} \cdot x = e \in H$ so the relation is reflexive. Notice that $x^{-1}y \in H$ if and only if $y^{-1}x \in H$, for these elements are multiplicative inverses of each other, and H is a group. Thus the relation is symmetric. Finally, suppose that $x, y, z \in G$, that x is related to y, and that y is related to z. Then $x^{-1} \cdot y \in H$ and $y^{-1} \cdot z \in H$. As a result, $[x^{-1} \cdot y] \cdot [y^{-1} \cdot z] = x^{-1} \cdot z \in H$. So we may conclude that x is related to z, and the relation is transitive.

Therefore we have an equivalence relation. The equivalence relation partitions the group G into pairwise disjoint subsets. What do these subsets look like?

If $x \in G$, then define $xH = \{x \cdot h : h \in H\}$. The set xH is called a *coset* of H. Notice the following properties:

(1) Elements of xH are Distinct: If $h, k \in H$, then $xh = xk$ implies $x^{-1}(xh) = x^{-1}(xk)$ hence $h = k$. So the elements of xH are distinct.

(2) If $a, b \in xH$, then $a\mathcal{R}b$: If $a, b \in xH$, then $a = xh$ and $b = xk$ for some $h, k \in H$. But then

$$a^{-1}b = (xh)^{-1}(xk) = h^{-1}x^{-1}xk = h^{-1}k \in H.$$

Thus a and b are related under $\mathcal{R}$.

(3) If $a \in xH$ and $a\mathcal{R}b$, then $b \in xH$: Now let $a \in xH$ and assume that $(a, b) \in \mathcal{R}$. Thus $a^{-1}b \in H$. So there is an element $h \in H$ such that $a = xh$ and there is another element $k \in H$ such that $a^{-1}b = k$. But then $b = ak = (xh)k = x(hk) \in xH$.

It follows from **(2)** and **(3)** that the equivalence classes induced by $\mathcal{R}$ are precisely the cosets of H. It follows from **(1)** that, if H has finitely many elements, then each coset has the same number of elements.

We let G/H denote the collection of cosets of H in G. With an additional condition on H (that H be a *normal subgroup*), G/H can actually be made into a group. We shall not explore that idea here.

Now we have:

Theorem 9.1.12 *Let* $G = \{g_1, \ldots, g_k\}$ *be a group with finitely many elements. Let* $H \subset G$ *be a subgroup with* m *elements. Then the integer* m *evenly divides the integer* k.

Proof: The group G partitions into the cosets of H. Each coset has m elements, and the cosets are of course pairwise disjoint. That means that m divides k. $\qquad\qquad\square$

EXAMPLE 9.1.13 Let a relation on the integers $\mathbb{Z}$ be defined by $x\mathcal{R}y$ if $y - x$ is evenly divisible by 6. This is an equivalence relation. There are six equivalence classes, namely

$$E_0, E_1, E_2, E_3, E_4, E_5.$$

Indeed,

$$
\begin{aligned}
E_0 &= \{\ldots, -12, -6, 0, 6, 12, \ldots\}, \\
E_1 &= \{\ldots, -11, -5, 1, 7, 13, \ldots\}, \\
E_2 &= \{\ldots, -10, -4, 2, 8, 14, \ldots\}, \\
E_3 &= \{\ldots, -9, -3, 3, 9, 15, \ldots\}, \\
E_4 &= \{\ldots, -8, -2, 4, 10, 16, \ldots\}, \\
E_5 &= \{\ldots, -7, -1, 5, 11, 17, \ldots\}.
\end{aligned}
$$

We add two equivalence classes as follows:

$$E_j + E_k = E_{j+k}.$$

For instance

$$E_3 + E_4 = E_7 = E_1.$$

You should check that this notion of addition is well defined (unambiguous). Also, the identity element is E_0, and each E_m has E_{-m} as its additive inverse. In sum, the collection of equivalence classes forms a group. This group is usually denoted by $\mathbb{Z}/6\mathbb{Z}$ or $\mathbb{Z}_6$ or $\mathbb{Z}/6$. It is a group having 6 elements; we call this a group of *order* 6. In general, the order of a group with finitely many elements is just the number of elements in the group.

Let H be the subset of $\mathbb{Z}_6$ consisting of E_0, E_2, E_4. Verify that this is a subgroup of order 3. Notice that 3 divides 6. The only other non-trivial subgroup is the one with elements E_0 and E_3. Notice that it has order 2.

The number 6 has no non-trivial divisors besides 2 and 3, and the group $\mathbb{Z}/6\mathbb{Z}$ has no other non-trivial subgroups.   ⋄

Of course there is nothing special about the number 6 in the last example. If k is any positive integer, then we may declare that $x\mathcal{R}y$ if $y - x$ is evenly divisible by k. The result is a group of order k denoted by $\mathbb{Z}/k\mathbb{Z}$ or $\mathbb{Z}_k$ or $\mathbb{Z}/k$. If the positive integer m evenly divides k ($k = m \cdot p$), then there will be one subgroup of order m and that group will consist of the elements $E_0, E_p, E_{2p}, \ldots, E_{(m-1)p}$. When the context is understood, we write the elements of $\mathbb{Z}_k$ as $0, 1, 2, \ldots, k - 1$. We say that we are doing *arithmetic modulo k* or *arithmetic mod k*.

EXAMPLE 9.1.14 Consider the set $\mathbb{Z}_4 \times \mathbb{Z}_4$. This set may be conveniently thought of as the set of ordered pairs (x, y) where $x, y \in \mathbb{Z}_4$. Define

$$(x, y) + (x', y') = (x + x', y + y'),$$

where the addition is performed according to the group law of $\mathbb{Z}_4$. Then $G = \mathbb{Z}_4 \times \mathbb{Z}_4$ so equipped is a group of order 16.

The number 4 divides the order of G, but now there is more than one subgroup having order (that is, number of elements) 4. One such subgroup is $H = \{(0,0), (1,0), (2,0), (3,0)\}$. Another is $K = \{(0,0), (0,1), (0,2), (0,3)\}$. Yet another is $L = \{(0,0), (2,0), (0,2), (2,2)\}$.   ⋄

Now let G be a group of finite order. Let g be a fixed element of G. Consider the set H of all "powers" of g: $g^1 = g, g^2 = g \cdot g, g^3 = g \cdot g \cdot g, \ldots$ as well as $g^{-1}, g^{-2} \equiv (g^{-1})^2 = g^{-1} \cdot g^{-1}, g^{-3} \equiv (g^{-1})^3, \ldots$. Of course $g^0 = e$. It is easy to see that H is a subgroup of G. We say that H is a *cyclic* group (subgroup) because it consists of powers of the single element g.

Let k be the order of H. Then k will be the least positive integer such that $g^k = e$. Since H is a subgroup of G, we see that $k = \operatorname{order} H$ must evenly divide $m = \operatorname{order} G$. It follows (provide the details as an exercise) that $g^m = e$. This conclusion is so important that we display it in a theorem:

Theorem 9.1.15 *Let G be a group of finite order m. If $g \in G$, then $g^m = e$.*

The examples we have presented raise a natural question. If G is a group of order m and if k evenly divides m, then does it follow that G has a subgroup of order k? In general the answer is "no." You are requested in the exercises to provide a counterexample. However, the following theorem of Sylow provides a positive answer in a large number of important instances:

Theorem 9.1.16 (Sylow) *Let G be a group of finite order, and let p be a prime. Suppose that j is a positive integer and that p^j evenly divides the order of G. Then G has a subgroup of order p^j. Indeed, it has subgroups of all orders p^ℓ, $0 \leq \ell \leq j$.*

Proof of a Special Case of the Sylow Theorems:

In this proof, we will use the standard notation $[G : H]$, for H a subgroup of G, to denote the quotient of the order of G by the order of H. In other words, $[G : H]$ is a positive integer.

Let G be a finite group, and suppose that the prime integer p divides the order n of G. We shall show that G has a subgroup of order p.

First, suppose that the integer m has the property that $g^m = e$ for every $g \in G$ (we call m an *exponent* for G). Let $b \in G$, $b \neq e$, and let H be the cyclic group generated by b (i.e., the collection of all powers of g). It can be checked that G/H forms a group. Then of course $b^m = 1$; hence m is an exponent for G/H. Thus the order of G/H divides a power of m. Because

$$[G : e] = [G : H] \cdot [H : e]$$

we may conclude therefore that the order of G divides a power of m.

Since p divides n, there an element $x \in G$ such that the period of x (i.e., the minimal power of x that equals e) is divisible by p. Let the period be ps for some integer s. Then $x^s \neq e$ and x^s has period p. Thus x^s generates a subgroup of order p. That is what was to be shown. □

We refer the interested reader to [LAN] or [HER] for a complete consideration of the Sylow theorems and their proofs.

We close with a few remarks about the concept of isomorphism of groups. Consider first an example. You are familiar with the group $\mathbb{Z}_2$. It is a group of order 2. Now suppose we consider a group G with two elements: $G = \{e, m\}$. The element e will be the group identity, and the rules of multiplication are

$$e \cdot m = m, \quad m \cdot e = m, \quad e \cdot e = e, \quad m \cdot m = e.$$

You can check for yourself that, with this binary operation, G is indeed a group—that is, it satisfies the axioms for a group.

On a formal level, the group $\mathbb{Z}_2$ and the group G are *different*. One has elements $0 \bmod 2$, $1 \bmod 2$, and the other has elements e, m. But in fact it turns out that they are the *same group*; they differ only in the sense that different names have been given to the elements. To see this, write out the group law for $\mathbb{Z}_2$:

$$0 + 1 = 1, \quad 1 + 0 = 1, \quad 0 + 0 = 0, \quad 1 + 1 = 0.$$

Actually do this on a piece of paper with a pencil. Now erase each occurrence of $+$ and replace it with $\cdot$. Also erase each occurrence of 0 and replace it by e and erase each occurrence of 1 and replace it by m. What results is the group law that we specified for the group G. This shows that $\mathbb{Z}_2$ and G are precisely the same group, with just different names for the elements and for the binary operation.

We have just seen an example of group isomorphism. Now we give a formal definition:

Definition 9.1.17 Let G and H be groups. A function $\phi : G \to H$ is said to be a *group isomorphism* if it has the following properties

1. The function ϕ is one-to-one and onto.

2. If $g_1, g_2 \in G$, then $\phi(g_1 \cdot g_2) = \phi(g_1) \cdot \phi(g_2)$.

We say that G and H are *isomorphic*, and we call the function ϕ an *isomorphism*.

It is the second condition of the definition that says, in effect, that both G and H have the same group law. Let us derive a few simple properties of group isomorphisms.

Proposition 9.1.18 Let $\phi : G \to H$ be an isomorphism of groups. If e_G is the group identity in G and e_H is the group identity in H, then $\phi(e_G) = e_H$.

Proof: We calculate that

$$\phi(e_G) = \phi(e_G \cdot e_G) = \phi(e_G) \cdot \phi(e_G).$$

The expressions that appear on the left and on the far right of this equation are elements of H. Multiplying both sides of this equation on the right by $[\phi(e_G)]^{-1}$, we find that

$$e_H = \phi(e_G).$$

That is what we wished to prove. $\square$

Proposition 9.1.19 Let $\phi : G \to H$ be a group isomorphism. If $g \in G$, then the group inverse, in the group H, of $\phi(g)$ is $\phi(g^{-1})$.

Proof: We may check that

$$\phi(g) \cdot \phi(g^{-1}) = \phi(g \cdot g^{-1}) = \phi(e_G) = e_H. \qquad \text{(by Proposition 9.1.18)}$$

Also

$$\phi(g^{-1}) \cdot \phi(g) = \phi(g^{-1} \cdot g) = \phi(e_G) = e_H. \qquad \text{(by Proposition 9.1.18)}$$

Thus $\phi(g^{-1})$ possesses the defining properties of the group inverse of $\phi(g)$. Since the group inverse of any group element is unique, our result follows. $\square$

The theory of groups has become a large and essential part of modern mathematics. It is also used in physics (in quantum mechanics, for instance), in engineering, and in theoretical computer science (for example, data compression theory uses group theory).

It is a classical result of basic group theory that all finite abelian groups have been classified. Indeed, it can be shown that any such group is a product (in the sense of set theory) of cyclic groups. One of the triumphs of twentieth-century mathematics is that all groups of finite order have been classified. This result is the product of the work of hundreds of mathematicians and will ultimately produce a book of several thousand pages.

See https://byjus.com/maths/group-theory for more about group theory.

Évariste Galois
1811–1832

Évariste Galois's father Nicholas Gabriel Galois and his mother Adelaide Marie Demante were both intelligent and well educated in philosophy, classical literature, and religion.

Galois had a difficult time in school. He had enrolled at the Lycée of Louis-le-Grand as a boarder in the 4th class on 6 October 1823. Even during his first term there was a minor rebellion and 40 pupils were expelled from the school.

February 1827 was a turning point in Galois' life. He enrolled in his first mathematics class, the class of M. Vernier. He quickly became absorbed in mathematics and his director of studies wrote, "It is the passion for mathematics which dominates him."

Tragedy was to strike Galois for on 2 July 1829 his father committed suicide. Galois was deeply affected by his father's death and it greatly influenced the direction his life was to take.

July 1830 saw a revolution. Charles 10th fled France. There was rioting in the streets of Paris and the director of École Normale, M. Guigniault, locked the students in to avoid them taking part. Galois tried to scale the wall to join the rioting but failed.

Two minor publications, an abstract in *Annales de Gergonne* (December 1830) and a letter on the teaching of science in the *Gazette des Écoles* (2 January 1831) were the last publications during his life.

The 14th of July was Bastille Day and Galois was arrested again. He was wearing the uniform of the Artillery of the National Guard, which was illegal. He was also carrying a loaded rifle, several pistols and a dagger. Galois was sent back to Sainte-Pélagie prison. While in prison he received a rejection of his memoir. While in Sainte-Pélagie prison Galois attempted to commit suicide by stabbing himself with a dagger but the other prisoners prevented him. While drunk in prison he poured out his soul: "Do you know what I lack my friend? I confide it only to you: it is someone whom I can love and love only in spirit. I have lost my father and no one has ever replaced him, do you hear me...?" In March 1832 a cholera epidemic swept Paris and prisoners, including Galois, were transferred to the pension Sieur Faultrier. There he apparently fell in love with Stephanie-Felice du Motel, the daughter of the resident physician. After he was released on 29 April Galois exchanged letters with Stephanie, and it is clear that she tried to distance herself from the affair.

Galois fought a duel with Perscheux d'Herbinville on 30 May, the reason for the duel not being clear but certainly linked with Stephanie.

Galois was wounded in the duel and was abandoned by d'Herbinville and his own seconds and found by a peasant. He died in Cochin hospital on 31 May and his funeral was held on 2 June. It was the focus for a Republican rally and riots followed which lasted for several days.

Augustin Louis Cauchy
1789–1857

Paris was a difficult place to live when Augustin-Louis Cauchy was a young child due to the political events surrounding the French Revolution. When he was four years old, his father, fearing for his life in Paris, moved his family to Arcueil. There things were hard. They soon returned to Paris and Cauchy's father was active in the education of young Augustin-Louis. Laplace and Lagrange were visitors at the Cauchy family home and Lagrange in particular seems to have taken an interest in young Cauchy's mathematical education.

In 1810, Cauchy took up his first job in Cherbourg to work on port facilities for Napoleon's English invasion fleet.

Back in Paris, Cauchy investigated symmetric functions and submitted a memoir on this topic in November 1812. An academic career was what Cauchy wanted and he applied for a post in the Bureau des Longitudes. He failed to obtain this post, Legendre being appointed. He also failed to be appointed to the geometry section of the Institute, the position going to Poinsot. His mathematical output remained strong and in 1814 he published the memoir on definite integrals that later became the basis of his theory of complex functions.

In 1817, when Biot left Paris for an expedition to the Shetland Islands in Scotland Cauchy filled his post at the Collége de France. There he lectured on methods of integration which he had discovered, but not published, earlier.

Cauchy did not have particularly good relations with other scientists. His staunchly Catholic views had him involved on the side of the Jesuits against the Académie des Sciences.

Political events in France meant that Cauchy was now required to swear an oath of allegiance to the new regime and when he failed to return to Paris to do so he lost all his positions there. In 1831, Cauchy went to Turin and after some time there he accepted an offer from the King of Piedmont of a chair of theoretical physics.

In 1843 Lacroix died and Cauchy became a candidate for his mathematics chair at the Collége de France. Liouville and Libri were also candidates. Cauchy should have easily been appointed on his mathematical abilities but his political and religious activities, such as support for the Jesuits, became crucial factors.

Another, rather silly, dispute this time with Duhamel clouded the last few years of Cauchy's life. This dispute was over a priority claim regarding a result on inelastic shocks. Duhamel argued with Cauchy's claim to have been the first to give the results in 1832. Poncelet referred to his own work of 1826 on the subject and Cauchy was shown to be wrong. However Cauchy was never one to admit he was wrong.

> ## Richard Dagobert Brauer
> ### 1901–1977
>
> Richard Brauer had an older brother Alfred Brauer, who also became a famous mathematician. Alfred Brauer was seven years older than Richard and of an age between the two brothers was Richard's sister Alice.
>
> Shortly after Brauer arrived in Königsberg, Knopp left to take up an appointment at Tübingen. The mathematics department at Königsberg was small, with two professors, Szegő and Reidemeister. This was the time when Brauer made his fundamental contribution to the algebraic theory of simple algebras.
>
> Brauer lost his position in Königsberg in the spring of 1933 after Hitler became Reichskanzler of Germany. Brauer was from a Jewish family so was dismissed from his post under the Nazi legislation which removed all Jewish university teachers from their posts. In November 1933, Brauer arrived to take up his appointment at the University of Kentucky, his wife and two sons following three months later. We should record that Alfred Brauer left Germany in 1939, but Brauer's sister Alice stayed behind and was murdered in a concentration camp by the Nazis.
>
> Following his year in Lexington, Brauer was appointed as Weyl's assistant at the Institute for Advanced Study in Princeton.
>
> A permanent post followed the two temporary posts when Brauer accepted an assistant professorship at the University of Toronto in Canada in the autumn of 1935.
>
> C. J. Nesbitt was Brauer's first doctoral student in Toronto and he described their relationship as doctoral student and supervisor: "Curiously, as thesis advisor, he did not suggest much preparatory reading or literature search. Instead we spent many hours exploring examples of the representation theory ideas that were evolving in his mind." It was in joint work with Nesbitt, published in 1937, that Brauer introduced the theory of blocks.
>
> Brauer was to spend the rest of his life working on the problem of classifying the finite simple groups. He died before the classification was complete but his work provided the framework of the classification which was completed only a few years later.
>
> Richard Brauer occupied an honoured position in the mathematical community, in which the respect due to a great mathematician was only one part. He was honoured as much by those who knew him for his deep humanity, understanding and humility; these were the attributes of a great man.

9.2   Euclidean and Non-Euclidean Geometry

Perhaps the first systematic, rigorous treatment of a subject in mathematics was Euclid's version of geometry (∼300 B.C.E.). While it is generally agreed that much of what Euclid recorded was due to scientists who came before Euclid, what is important about Euclid's contribution is the *manner in which he organized and recorded the subject*.

Euclid's geometry has five undefinable terms. These are *point*, *line*, the phrase *"lie on,"* the concept of *between*, and *congruent*. The notion of circle can in fact be defined in terms of these undefinables (i.e., the term "circle" itelf is *not* an undefinable):

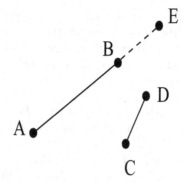

FIGURE 9.1
Euclid's Axiom 2.

Definition 9.2.1 Let two points O and A be given. The set of all points P such that the segment OP is congruent to segment OA is called a *circle* with center O. Each of the segments OP is called a *radius* of the circle.

Euclid's geometry has five axioms (see [GRE]). They are as follows:

P1 Through any pair of distinct points, there passes a line.

P2 For each segment $\overline{AB}$ and each segment $\overline{CD}$, there is a unique point E (on the line determined by A and B) such that B is between A and E and the segment CD is congruent to BE. See Figure 9.1.

P3 For each point C and each point A distinct from C, there exists a circle with center C and radius CA.

P4 All right angles are congruent.

P5 For each line ℓ and each point P that does not lie on ℓ, there is a unique line m through P such that m is parallel to ℓ.

We have not been as careful here as would be necessary in a thorough development of geometry. In particular, you should note that we have not provided all the necessary definitions. As an exercise, you may wish to study these axioms, determine which terms require definition, and provide those definitions. One important instance of our omissions occurs in Axiom **P5**: we did not first define "parallel." One working definition of two lines being parallel is that they do not intersect. Another working definition is that there exists a transversal line that is perpendicular to both lines. You should also think about the definitions of "right angle" and "congruent segments" and "perpendicular." A thorough discussion of all these ideas appears in [GRE].

At this time you should review the discussion of models, consistency, and independence that appears in the Appendix. We know that the five axioms of

Euclid's geometry are consistent because there is a *model* that satisfies these axioms. That model is the usual geometry that we learn in high school. It consists of the lines that we draw with a ruler and the circles that we draw with a compass. [We could digress at this point into a discussion of what it means to verify that said lines and circles satisfy the axioms, but instead we refer the reader to [GRE].]

For nearly 2000 years, scholars wrestled with the question of whether the first four axioms could be used to prove the fifth, the so-called parallel postulate. In other words, is **P5** *independent* of the other four axioms? In 1829, Nikolai Ivanovich Lobachevski (1793–1856) published the construction of a geometry that satisfies Euclid's first four axioms, but such that the parallel postulate fails. János Bolyai (1775–1856) constructed such a geometry at roughly the same time, but his ideas were published somewhat later. Carl Friedrich Gauss (1777–1855) claimed in a private communication to Bolyai's father Farkas Bolyai that he had made these discoveries several years earlier, but had not published for a variety of enigmatic reasons. Gauss expressed his thoughts on the matter, in his letter to János Bolyai, as follows:

> I am exceedingly sorry that I have failed to avail myself of our former great proximity to learn more of your work on the foundations of geometry; it surely would have saved me much useless effort and given me more peace, than one of my disposition can enjoy so long as so much is left to consider in a method of this kind. I have myself made much progress in this matter (though my other heterogeneous occupations have left me but little time for this purpose); though the course which you assure me you have reached as to the questioning of the truth of geometry. It is true that I have found much which many would accept as proof, but which in my estimation proves *nothing*, for instance, if it could be shown that a rectilinear triangle is possible, whose area is greater than that of any given surface, then I could rigorously establish the whole of geometry. Now most people, no doubt, would grant this as an axiom, but not I; it is conceivable that, however distant apart the vertices of the triangle might be chosen, this area might yet always be below a certain limit. I have found several other such theorems, but none of them satisfies me.

A delightful account of the history of these matters appears in [GRE].

In a moment we shall present a version of the non-Euclidean geometry of Bolyai and Lobachevski. First let us consider what it means. We have already acknowledged that there is a model for the five axioms of Euclid's geometry. In other words, in the classical rectilinear geometry of high school, Axioms **P1**, **P2**, **P3**, **P4**, and **P5** are true. So **P1**, **P2**, **P3**, **P4**, and **P5** are *consistent*. Now we are saying that there is a model for **P1**, **P2**, **P3**, **P4**, and $\sim$ **P5**. This says that **P1**, **P2**, **P3**, **P4**, and $\sim$ **P5** are consistent. In other words, because of the standard Euclidean model, **P5** is consistent with the other four axioms; and because of the new model that we are about to present it is the case that $\sim$ **P5** is consistent with the other four axioms. This means that neither **P5** nor $\sim$ **P5** could be derived from the other four axioms. In other words, the fifth axiom is independent of the first four.

Gauss claimed that he did not publish his construction of non-Euclidean geometry in part because nobody else would understand its import. It is certainly the case that the formal ideas of model, consistency, and independence had not yet been developed at that time. In this text we have taken a little care to set the stage for non-Euclidean geometry.

There are many versions of non-Euclidean geometry. The theories of Beltrami–Klein and Poincaré figure prominently among these. It is a theorem that all two-dimensional non-Euclidean geometries are congruent (see [GRE]). However, some features of the geometry are clearer in one model and other features are clearer in another model. Our decision to present Poincaré's model (below) is one based only on taste. [We note in passing that all models for *Euclidean geometry* are congruent as well. This is proved by way of the cartesian coordinate system. See [GRE] for details.]

At the risk of overstating the case, let us repeat what we are about to do. We shall not present any new axioms or new ideas. We shall instead present a new notion of "line" that satisfies **P1**, **P2**, **P4** (Axiom **P3** is irrelevant for the present discussion) but does not satisfy **P5**. The notion of "point" will remain the same. These new lines cannot be perfectly arbitrary objects. They must have a certain Euclidean-like structure; that is, they must satisfy the first three axioms. But they will not satisfy the fifth.

János Bolyai (1775–1856)

Farkas Bolyai always wanted his son János to be a mathematician. Until János was nine years old, the best students from the Marosvásárhely College taught him all the usual school subjects except mathematics, which he was taught by his father. By the time Bolyai was 13, he had mastered the calculus.

When János graduated from Marosvásárhely College on 30 June 1817, it was not clear how he might obtain a good mathematical education. The decision that János would study military engineering at the Academy of Engineering at Vienna was not taken without a lot of heartache. János next was admitted to the Academy in Vienna.

He studied at the Royal Engineering College in Vienna from 1818 to 1822, completing the seven year course in four years.

In September 1823 he entered the army engineering corps. He spent a total of 11 years in military service and was reputed to be the best swordsman and dancer in the Austro-Hungarian Imperial Army.

Around 1820, when he was still studying in Vienna, Bolyai began trying to replace Euclid's parallel axiom with another axiom which could be deduced from the others. On 3 November 1823 he wrote to his father that he had

... created a new, another world out of nothing ...

but he still added a few lines later that it was not created yet. By 1824, however, there is evidence to suggest that he had developed most of what would appear in his treatise as a complete system of non-Euclidean geometry.

By 1830 Farkas had come to understand the full significance of what his son had accomplished and strongly encouraged him to write up the work for publication. By 20 June 1831 the Appendix had been published. Farkas Bolyai sent a reprint to Gauss. Gauss wrote to Farkas Bolyai

To praise it would amount to praising myself. For the entire content of the work ... coincides almost exactly with my own meditations which have occupied my mind for the past thirty or thirty-five years.

The discovery that Gauss had anticipated much of his work, however, greatly upset Bolyai who took it as a severe blow. His health began to deteriorate and he found it difficult to carry out his military duties.

Certainly Bolyai continued to develop mathematical theories. His one major undertaking, to attempt to develop all of mathematics based on axiom systems, was begun in 1834.

In 1848, Bolyai discovered that Lobachevsky had published a similar piece of work in 1829. János studied Lobachevsky's work carefully. The work stirred a real storm in his soul.

János Bolyai gave up working on mathematics in his last years and instead tried to construct a theory of all knowledge. In 1945 the Hungarian university in Cluj was named after him.

The resulting geometry will have a number of peculiar features—peculiar, that is, when viewed from our familiar Euclidean perspective. One of the

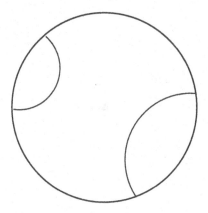

FIGURE 9.2
Some Poincaré lines.

reasons that Gauss invoked for not publishing his ideas about non-Euclidean geometry is that he expected that his fellow mathematicians and philosophers would find this all to be an artificial construct of the mind: just a game with ideas that has nothing to do with the real world.

Many times it requires a great deal of time for mathematical ideas to find applications in the real world. It turns out that the model for non-Euclidean geometry that we shall describe, often called *hyperbolic geometry*, is the key to determining the structure of four-dimensional manifolds or surfaces. Four-dimensional manifolds are in turn the language for studying general relativity. Thus, after nearly two centuries, we find that non-Euclidean geometry *is* important in the world around us. However, hyperbolic geometry is more difficult for us to perceive intuitively than standard Euclidean geometry.

The two-dimensional geometry that we now describe takes place not in the infinitely extendible plane but rather inside the unit circle C. Any standard Euclidean point *in the interior* of C is a point of our geometry. We declare a (Poincaré) *line* to be any circular arc that begins and ends on C and is perpendicular to C at both endpoints *or* any standard straight line segment passing through the origin and touching C at both ends. Figures 9.2 and 9.3 illustrate our notion of "line" in this new geometry.

Notice that the second type of line in our geometry—the line segment through the origin—is also perpendicular to C at its endpoints. This is the key geometric fact that characterizes Poincaré lines. Also note that, when we speak of two circular arcs being perpendicular at a point P, then we mean, of course, that their tangent lines are perpendicular at P. An equivalent, but more elementary, way to think about this idea is that the *radii* of the two circular arcs are perpendicular at P: see Figure 9.4.

In what follows we shall engage in an extensive discussion of the artifacts of non-Euclidean geometry. Along the way, we shall describe a variety of

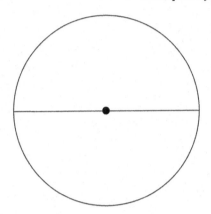

FIGURE 9.3
Another Poincaré line.

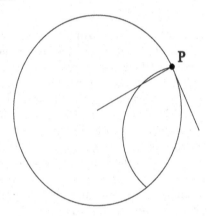

FIGURE 9.4
Perpendicular radii.

geometric constructions. Since this is only an introduction, we shall not provide complete proofs of why the constructions work. We refer the reader to [GRE] for details.

It is not immediately clear that the first type of line even exists. Here is how we can construct them: let x and y be arbitrary points of C that are not on the same diameter. Draw the tangent lines to C at each of these points. These tangents intersect at a unique point z. The circle with center z and passing through x and y will be perpendicular to C at both points of intersection (Figure 9.5). Thus the arc of this circle that lies inside C will be a line in our new geometry.

Why is it true that, through any two points inside C, there passes a line? More fundamentally, why is it true that, through any single point, there passes

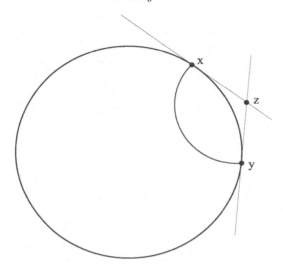

FIGURE 9.5
A line in the new geometry.

a line? In order to answer these questions most expeditiously, we need to introduce the concept of *reflection in a circle* in the circle C. Let P be a point inside C. Let O denote the center of C. The reflection of P is the unique point $\widetilde{P}$ on the ray $\overrightarrow{OP}$ such that $|OP| \cdot |O\widetilde{P}| = 1$. Refer to Figure 9.6.

Now here is the interesting fact: The circle with center $\widetilde{P}$ and passing through P will be orthogonal to C at the points of intersection. Thus the arc of that circle that lies inside C will be a line in our geometry, and it will pass through P. Look again at Figure 9.6. This answers the request for a line through any single given point.

To answer the second question (and establish Axiom **P1**), it is convenient to construct an isomorphism with another model for non-Euclidean geometry—the Beltrami–Klein model. To do this, we use the so-called stereographic projection. As Figure 9.7 suggests, we may project a point P of the Poincaré disc up to a point $\pi(P)$ in the unit sphere by moving along a segment in space that connects the planar point to the north pole N of the sphere. We then project the spherical image vertically down to a point P' in the plane again.

The result of the composition of these two projections is a new planar geometry in which the role of "line" is played by classical straight segments that lie inside a (slightly smaller) circle. Now if we want to know whether two points P and Q in Poincaré's geometry have a (Poincaré) line passing through them, we project the two points up into the sphere and vertically down into the plane again to obtain two points P' and Q' in the Beltrami–Klein geometry. We draw an *ordinary segment S* in the Beltrami–Klein model, passing through P' and Q', project it up vertically into the sphere, and then

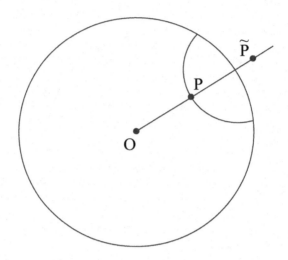

FIGURE 9.6
Reflection of the point P.

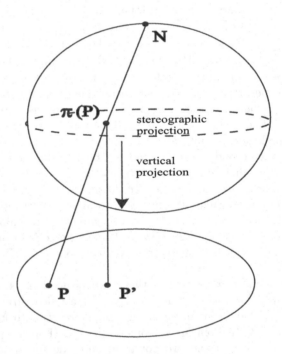

FIGURE 9.7
Projection to the unit sphere.

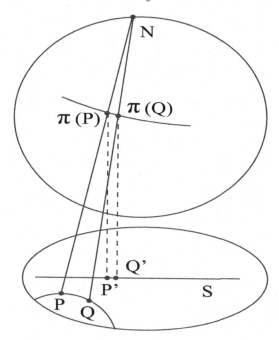

FIGURE 9.8
A Poincaré line by way of the inverse stereographic projection.

apply the inverse stereographic projection to push the circular arc back into the plane. This gives a Poincaré line that passes through the original two points P and Q. Refer to Figure 9.8. Thus we find that Axiom **P1** is satisfied by the (Poincaré) points and (Poincaré) lines in Poincaré geometry.

In Poincaré geometry, the angle between two Poincaré lines at a point of intersection is just declared to be the angle between two classical circular arcs which is defined to be the classical Euclidean angle between the tangent lines at that point. Thus Axiom **P4** is satisfied.

In order to satisfy Axiom P2, we must reconsider our notion of distance. For look at the situation shown in Figure 9.9. If our idea of distance is the ordinary Euclidean idea (measured along a curve with a tape measure for instance), then we would be unable to construct the required point E for these points A, B, C, D. We thus need to modify the notion of distance so that points near the boundary are further apart than points near the origin. With this goal in mind, consider a fixed ruler that has length one inch at the origin; the ruler will change scale when it is moved nearer to the boundary of the disc. In other words, our notion of length is dependent on the base point at which we are standing.

The ruler will have length one foot if it is moved sufficiently near the boundary and length one yard if it is moved even nearer. If it is very near

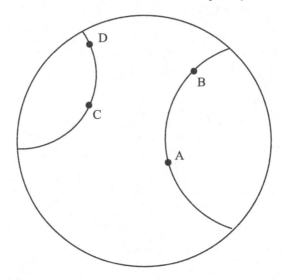

FIGURE 9.9
A new notion of distance

indeed to the boundary, then that same yardstick will have length one mile. As the base point tends to the boundary, the length of the ruler will approach infinity.

How can this be? We define the *Poincaré distance* between two points x and y in the unit disc as follows: let P and Q be the endpoints of the Poincaré line through x and y lying on the circle C as in Figure 9.10. We define

$$\rho(x,y) = \left| \log\left(\frac{|xQ||yP|}{|yQ||xP|} \right) \right|.$$

Here $|xP|$ stands for the ordinary Euclidean distance of x to P and similarly for the other expressions.

In order to get a feeling for this notion of distance, we calculate $\rho(O,X)$, where O is the center of the circle and $X = (r,0)$ with $0 < r < 1$. Of course the Poincaré line through O and X is just the ordinary horizontal diameter segment of C. Now P is the point $(-1,0)$ and Q is the point $(1,0)$. Then

$$
\begin{aligned}
|OQ| &= 1 \\
|XP| &= 1+r \\
|OP| &= 1 \\
|XQ| &= 1-r.
\end{aligned}
$$

As a result,

$$\rho(O,X) = \log\left(\frac{1+r}{1-r} \right).$$

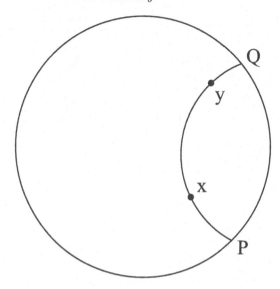

FIGURE 9.10
Endpoints of a Poincaré segment.

Notice that, as r tends to 1 from the left, the value of $\rho(O, X)$ tends to infinity. In particular, it must be that points out near 1 are getting further and further from the origin and also further and further apart.

An interesting feature of the notion of distance that we have introduced is that it makes the unit disc much like the Euclidean plane equipped with the ordinary Euclidean metric: that is, the unit disc with the Poincaré metric has infinite extent in all directions. It is also the case that, just as lines are the curves of shortest length in the Euclidean metric in the plane, so it is that Poincaré lines are the curves of shortest length for the Poincaré metric in the unit disc.

Now return to Figure 9.9. The distance along the curve through A and B and out to the boundary is infinity (Figure 9.11). So there is plenty of room to construct E as in Axiom **P2** (Figure 9.12).

Let us cut this preliminary discussion short and finally consider the crucial Axiom **P5**. First, what does it mean for two lines to be parallel? By definition, they are parallel if they do not intersect. Figure 9.13 exhibits some parallel lines. Now look at Figure 9.14: it exhibits a line ℓ and a point x not on that line. Figure 9.15 shows two distinct Poincaré lines passing through x that are parallel to ℓ. In fact there are infinitely many such lines! Thus we see that Axiom **P5** fails for Poincaré geometry.

Space limitations prevent us from exploring this new hyperbolic (non-Euclidean) geometry to any depth. It contains many beautiful subtleties. For example, all triangles in non-Euclidean geometry have angles summing to less than 180°. There are no rectangles (quadrilaterals with four right angles) in

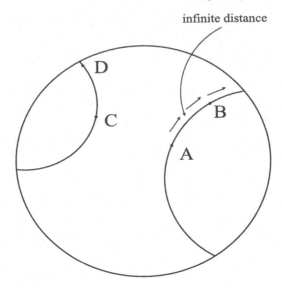

FIGURE 9.11
Infinite distance to the boundary.

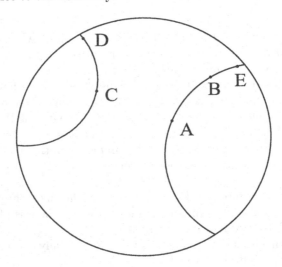

FIGURE 9.12
Euclid's second axiom in the Poincaré setting.

non-Euclidean geometry. Perhaps most surprising is that two similar triangles must be congruent: that is, if their angles are the same, then their side lengths must also be the same. Thus, in particular, photography would not work very well in non-Euclidean geometry: a lens that shrunk or expanded objects would

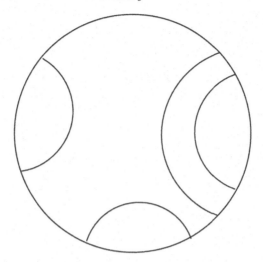

FIGURE 9.13
Parallel lines in the Poincaré geometry.

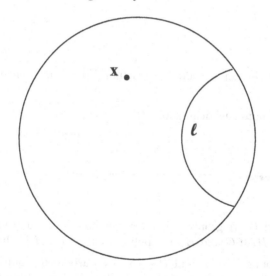

FIGURE 9.14
Failure of the parallel postulate in Poincaré geometry.

perforce have to change angles and therefore would also distort the object being photographed.

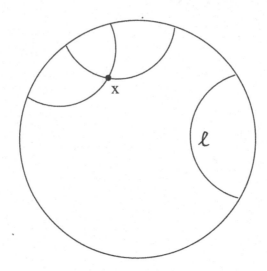

FIGURE 9.15
Another look at the failure of the parallel postulate.

See

https://www.britannica.com/science/non-Euclidean-geometry

for more about non-Euclidean geometry.

Exercises

1. If G and H are groups, then describe how to put a group structure on $G \times H$. If G and H are abelian then is $G \times H$ abelian?

2. The groups $G = \mathbb{Z}_4$ and $H = \mathbb{Z}_2 \times \mathbb{Z}_2$ are both groups of order 4. But they are not the same group. That is, they are not isomorphic. Give a proof of this last statement.

3. Describe all possible groups of order 2.

4. Describe all possible groups of order 3.

5. Describe all possible groups of order 4.

6. Independent of having a model for non-Euclidean geometry, we could consider a (theoretical) geometry based on Euclid's axioms **P1–P4** and a fifth axiom that says:

Q5 There exists a line ℓ and a point P not on the line such that at least two distinct lines parallel to ℓ pass through P.

Show that, in this geometry, there is a triangle the sum of whose angles is less than 180°.

7. Notice that **Q5** in the last exercise is precisely the negation of the parallel postulate **P5** in the text. But we might ask if it then follows logically that for *every* line m and *every* point S not on m there are at least two distinct lines through S that are parallel to m. Prove that this is so.

8. Refer to Exercises 9.6 and 9.7. Prove that non-Euclidean geometry contains no rectangles.

9. Let G be a group and H a subgroup. We say that H is *normal* if, whenever $g \in G$ and $h \in H$, then $g^{-1}hg \in H$. Prove that, when H is a normal subgroup, then the collection of cosets G/H forms a group. Your group operation should be as follows: if xH and yH are cosets, then their product is defined to be xyH. You must prove that this is well defined and gives an associative binary operation on G/H. You must also show that there is a group identity and that every element has an inverse.

10. Let S be a set and let • be an operation on that set. Thus if s, t are elements of S, then $s \bullet t$ is another element of S. Let $T \subset S$. We say that the operation • is *closed* on T if $s \bullet t$ lies in T whenever $s, t \in T$.

Let S be the integers and let $T = \{\ldots, -5, -1, 3, 7, 11, \ldots\}$. Is T closed under addition? Is T closed under multiplication?

Let S be the integers and let $T = \{\ldots, -4, -2, 0, 2, 4, \ldots\}$. Is T closed under addition? Is T closed under multiplication?

11. In view of what you learned in the Appendix and Section 9.2, discuss the validity of the following statement: "If Euclidean geometry is consistent, then hyperbolic geometry is consistent."

12. A group G is called *cyclic* if there is an element $x \in G$ such that every element of G has the form x^k for some fixed element $x \in G$. Here $x^k \equiv x \cdot x \cdots x$, the product of k copies of x, when k is a positive integer; $x^k \equiv (x^{-1})^{|k|}$ when k is a negative integer; and $x^0 = e$. Give an example of a cyclic group with infinitely many elements (that is, a cyclic group of infinite order). Give an example of a cyclic group with finitely many elements (that is, a cyclic group of finite order).

13. Refer to Exercise 9.9 and 9.12 for terminology. If G is a cyclic group and H is a subgroup, then prove that G/H is a cyclic group.

14. Let $G = \mathbb{Z}_6 \times \mathbb{Z}_9$. Identify all the subgroups that have order a power of 3.

* **15.** Construct a hyperbolic trigonometry based on the hyperbolic geometry of the unit circle.

16. As noted in the text, the stereographic projection can be used to give an isomorphism between the Poincaré version of non-Euclidean geometry and the Beltrami–Klein version. Discuss Axioms **P1**–**P5** as well as **Q5** (Exercise 9.6 above) for the Beltrami–Klein model.

17. The complex function
$$\phi(z) = i\frac{1-z}{1+z}$$
maps the unit disc (the interior of the unit circle C) one-to-one and onto the upper half plane $U \equiv \{x + iy : y > 0\}$. This mapping is infinitesimally angle preserving (or *conformal*). Thus it induces a non-Euclidean geometry on the upper half plane. Describe this geometry. What are the points? What are the lines? What is the notion of distance?

18. In calculus we learn to calculate the length of a curve $\gamma(t), 0 \le t \le 1$, according to the formula
$$\ell(\gamma) = \int_0^1 |\gamma'(t)|\, dt.$$

Here $|\gamma'(t)|$ is the standard Euclidean length of the tangent vector $\gamma'(t)$.

Riemann's approach to non-Euclidean geometry was to allow the method of measuring the length of a tangent vector to vary from point to point. After all, $\gamma'(t_1)$ is a tangent vector to the curve *at the point* $\gamma(t_1)$ and $\gamma'(t_2)$ is a tangent vector to the curve *at the point* $\gamma(t_2)$, and it is plausible that length could be measured differently at these two points.

Now if $\gamma(t), 0 \le t \le 1$, is a curve in the unit disc, let us define the *Poincaré length* of the tangent vector $\gamma'(t)$ to be
$$\|\gamma'(t)\|_{P,\gamma(t)} = \frac{2|\gamma'(t)|}{1 - |\gamma(t)|^2}.$$

As before, $|\ \ |$ denotes Euclidean length. Notice that we are dilating our method of measuring length according to how close the point is to the boundary.

Fix $0 < r < 1$. Use the Poincaré length of a tangent vector to calculate the length of the curve $\gamma(t) = rt, 0 \le t \le 1$. Compare your result with the calculation of the Poincaré distance of 0 to $(r, 0)$ that was given in the text.

19. Refer to Exercise 9.18 for terminology. Consider geometry on the Euclidean plane. Suppose that $\gamma(t) = (\gamma_1(t), \gamma_2(t))$, $0 \leq t \leq 1$, is a curve. Define the length of a tangent vector $\gamma'(t)$ at $\gamma(t)$ to be

$$\|\gamma'(t)\|_{\gamma(t)} = \max\{|\gamma_1'(t)|, |\gamma_2'(t)|\}.$$

In this geometry, what is the distance from $(0,0)$ to $(1,1)$? What is the curve of least length connecting these two points?

20. The Euclidean space $\mathbb{R}^3$ can be equipped with a group structure according to the law $(x, y, z) + (x', y', z') = (x + x', y + y', z + z')$. In this structure, the element $(0,0,0)$ is the identity and the inverse of (x, y, z) is $(-x, -y, -z)$. However, this is not the only group structure on $\mathbb{R}^3$. Devise another group structure that is non-commutative. (**Hint:** The group law will be expressed as quadratic polynomials in the coordinates. Consider a subgroup of the 3×3 matrices under matrix multiplication.)

21. Consider the group of all 2×2 matrices with entries that are elements of $\mathbb{Z}_3$. How many elements does this group have? Give an example of a subgroup of order 9. Give an example of a subgroup of order 27. Give an example of a subgroup of order 3.

* 22. Give an example of a group G of finite order m and an integer k that divides m such that G does *not* have a subgroup of order k. [**Hint:** The group must be non-abelian. This problem is tricky. The least k, m that will work are 6 and 12.]

23. Let G and H be isomorphic groups. Call the isomorphism ϕ. Suppose that $g \in G$ and that $g^k = e_G$ for some positive integer k. Prove that $\phi(g)$ has the property that $[\phi(g)]^k = e_H$.

24. Refer to Exercise 9.9 for terminology. Let G and H be isomorphic groups, with isomorphism ϕ. Let K be a normal subgroup of G. Prove that the image of K under ϕ is a normal subgroup of H.

25. Let a be a complex number of modulus less than 1. Define $\phi_a(z) = (z - a)/(1 - \bar{a}z)$. Prove that ϕ_a maps the interior of the unit circle to the interior of the unit circle. Indeed, ϕ_a maps the unit circle itself to the unit circle. Prove that ϕ_a maps lines in Poincaré geometry to lines in Poincaré geometry. In particular, a circular arc inside the unit circle that is perpendicular to the unit circle is mapped by ϕ_a either to another such circular arc, or to a diameter of the circle.

26. Let p and q be points on the unit sphere. Describe an algorithm for finding the great circle that passes through p and q. Under what circumstances can you be sure that this circle is unique?

27. Refer to Exercise 9.25 for terminology. For $0 < \theta \leq 2\pi$, define a map

$$\rho_\theta(z) = e^{i\theta} \cdot z.$$

Then ρ_θ is the rotation of the unit disc in the complex plane through an angle of θ. Prove this statement, and prove that ρ_θ is one-to-one and onto from the disc to itself.

Prove that the collection of all maps of the form

$$\rho_\theta \circ \phi_a$$

forms a group. This object is known as the "group of conformal self-maps of the unit disc."

28. Refer to Exercises 9.25 and 9.27 for terminology. Prove that, if there are two points p, q in the disc such that a mapping $\tau = \rho_\theta \circ \phi_a$ satisfies $\tau(p) = p$ and $\tau(q) = q$, then in fact $\tau(z) \equiv z$ for all $z \in D$.

29. In the standard geometry of the Euclidean plane, the curves of shortest distance are straight lines: the least distance between two planar points is realized by the straight Euclidean line between those points. The Poincaré lines play the same role in the Poincaré geometry of the unit disc.

    Conversely, in classical Euclidean geometry, we know that two planar points uniquely determine a straight line. What is the correct analog of "two" for the Poincaré lines in the disc?

30. Let ℓ be a Poincaré line in the unit disc and p a point that is not on that line. Is there a unique Poincaré line m through p that is perpendicular to ℓ?

APPENDIX: Axiomatics

A1 Axioms of Set Theory

In a rigorous treatment of set theory such as one encounters in an advanced course in logic, there is a detailed consideration of the *axioms* of set theory. As indicated in Chapter 1, this is part and parcel of the way that mathematics is done "for the record." It is the way that we insure that no logical inconsistencies will arise in our future work, and it is the way that we lay the ground rules for our subject.

The present book constitutes your first exposure to rigorous mathematical thinking, and it would be somewhat overwhelming for you to have to wade through a thorough treatment of the axioms of set theory at this time. We have instead been studying a "naive" treatment of set theory. Nonetheless, we should like now briefly to treat the formal axioms and to discuss some of the more significant axioms that arise frequently in advanced mathematics.

The Axioms of Zermelo–Fraenkel Set Theory

Axiom of Extensionality

$$\big[\forall x, (x \in A \text{ iff } x \in B)\big] \Rightarrow (A = B)$$

Sum Axiom Given a collection $\mathcal{S}$ of sets:

$$\exists C \ \forall x, \left(x \in C \text{ iff } \exists B, (x \in B \land B \in \mathcal{S}) \right)$$

Power Set Axiom Given a set A:

$$\exists B \ \forall C, (C \in B \text{ iff } C \subset A)$$

Axiom of Regularity

$$A \neq \emptyset \Rightarrow \exists x, \big[x \in A \land \forall y, (y \in x \Rightarrow y \notin A)\big]$$

Axiom for Cardinals If A is a set, then let $\mathcal{K}(A)$ denote the set of all sets that are set-theoretically isomorphic to A. The set $\mathcal{K}(A)$ is sometimes called the *cardinality of A*, or the *cardinal number* corresponding to A. Then

$$\mathcal{K}(A) = \mathcal{K}(B) \text{ iff } \mathrm{card}(A) \equiv \mathrm{card}(B),$$

where $\equiv$ means set-theoretic isomorphism.

Axiom of Infinity

$$\exists A, \left(\emptyset \in A \wedge \forall B, [B \in A \Rightarrow B \cup \{B\} \in A] \right)$$

Axiom Schema of Replacement Let $P(x, y)$ be a property of x, y. If

$$\forall x \, \forall y \, \forall z, \left[\left\{ x \in A \wedge P(x, y) \wedge P(x, z) \right\} \Rightarrow \left\{ \mathbf{y} = \mathbf{z} \right\} \right]$$

then

$$\exists B \, \forall y, [y \in B \Leftrightarrow \exists x, (x \in A \wedge P(x, y))].$$

Axiom of Choice For any set A, there is a function $f : \mathcal{P}(A) \to A$ such that, for any non-empty subset B of A, we have $f(B) \in B$.

The advantage of the way that we have stated the axioms here is that the actual statements do not involve English words (which are subject to misunderstanding). The statements are also brief. But it requires some effort to understand what they say. To aid in this process, we now give a brief, informal description of what each axiom says.

The Axiom of Extensionality mandates that two sets are equal precisely when they have the same elements. In our treatment in Section 3.1, we took this property as a definition.

The Sum Axiom specifies that if we are given a collection of sets $\mathcal{S} = \{S_\alpha\}_{\alpha \in A}$, then the union set $C = \cup_{\alpha \in A} S_\alpha$ exists. Again, in the interest of simplicity we treated this idea in Chapter 3 with a definition. Lurking in the background here is the fact (to be explored briefly at the end of this section) that we cannot allow the existence of sets that are too large. If we do, then certain paradoxes result. Thus certain of our set theory axioms specify the ways in which we are allowed to form new sets.

The Power Set Axiom says that if we are given a set A, then its power set $\mathcal{P}(A)$ also exists. The comments in the last paragraph about creating new sets also apply here.

The Axiom of Regularity is slightly more subtle. Consider the set S of all sets that can be described with fewer than 50 words. Then S is an element of itself. Such considerations can lead to nasty paradoxes (see Russell's paradox at the end of this section). For this reason, we want to rule out sets that are elements of themselves. As an exercise, you may try to prove, using the Axiom of Regularity, that if A is a set, then $A \notin A$.

The Axiom for Cardinals addresses another existence problem. Given a set S, we wish to consider the set of sets each of which has the same cardinality as S. This is an equivalence class (with the equivalence relation being set-theoretic isomorphism), but we do not know in advance that the equivalence class exists. The Axiom for Cardinals mandates just that. Again, this is one of our allowed methods for creating new (possibly large) sets. Some versions of set theory disallow extremely large sets by not including the Axiom for Cardinals.

The Axiom of Infinity specifies the existence of a set that is set-theoretically isomorphic to a proper subset of itself. We know from Section 4.5 that this is equivalent to postulating the existence of an infinite set. In fact, our formulation of the Axiom of Infinity is very closely related to the classical construction of ordinal numbers and, more specifically, of the natural numbers themselves. See the discussion in Chapter 5.

The Axiom of Replacement is perhaps the most technical axiom, but it is also the one that is used most often. It is our device for passing from the general to the specific. It is our means of specifying a set as the collection of objects that satisfy a certain property. The first line of the axiom specifies that P is a property that can be treated. The second line specifies that there exists a set B that is the set of elements that satisfy property P.

The Axiom of Choice is perhaps the axiom that has been the richest source of ideas, and also of confusion, in all of set theory. Its intuitive statement is simplicity itself: Let S be a set. Then there is a function $f : \mathcal{P}(S) \to S$ that assigns to each non-empty subset of S an element of itself. While quite a plausible axiom, it is a metatheorem of mathematical logic that it is impossible to specify the function that assigns to each non-empty subset of the reals an element of itself. By contrast, with $S = \mathbb{N}$, the natural numbers, we can do it: if $A \subset \mathbb{N}$, then let $f(A)$ be the least element of A. We have proved in our treatment of the natural numbers in Section 5.1 that every subset of the natural numbers does indeed *have* a least element. So this choice function f makes good sense and does the job. Think for a moment why a function defined in this way will not work for the real numbers.

We shall spend the next section discussing the Axiom of Choice and its consequences.

We close this section with a brief mention of Russell's paradox. There are many paradoxes that fall under this rubric, and Russell's was not the first. A nice discussion of these matters appears in [SUP, Ch.1]. See also [JEC].

Russell's paradox: Let S be the set of all sets which are not elements of themselves (most sets fit this description, although we gave an example of a set which *is* an element of itself in the discussion of the Axiom of Regularity). Is S an element of itself?

If S is an element of itself then, by definition, S is not an element of itself. On the other hand, if S is not an element of itself then it must be an element of itself (since S has as elements *all* sets which are not elements of themselves).

Thus S can neither be an element of itself nor not be an element of itself.

Russell's paradox was first communicated to Frege just weeks before the latter's book on the foundations of set theory was to appear in print. The paradox called into question Frege's, and everyone else's, approach to set theory. It is because of Russell's paradox that modern versions of set theory, such as that given by the eight axioms above, set very careful restrictions on which sets we may construct and consider. In particular, the set S of Russell's paradox is too large to exist in the version of Zermelo–Fraenkel set theory that was discussed earlier in this section. The concept of "class" has been developed to deal with objects that are too large to be considered a part of set theory. See [FRA] or [COH] for a discussion of classes.

We close this section by summarizing the discussion: In a formal treatment of set theory, one begins with the eight axioms specified above and with the empty set and systematically develops the properties of sets. This is done, for instance, in [SUP]. In the present text, we have given an intuitive treatment of set theory that parallels the classical, rigorous treatment but is more accessible.

Bertrand Arthur William Russell
1872–1970

Russell was born the grandson of Lord John Russell, who had twice served as Prime Minister under Queen Victoria. Following the death of his mother (in 1874) and of his father (in 1876), Russell and his brother went to live with their grandparents.

Although elected to the Royal Society in 1908, Russell's career at Trinity appeared to come to an end in 1916 when he was convicted and fined for anti-war activities. He was dismissed from the College as a result of the conviction. Married four times and notorious for his many affairs, Russell also ran unsuccessfully for Parliament, in 1907, 1922, and 1923.

While teaching in the United States in the late 1930s, Russell was offered a teaching appointment at City College, New York. The appointment was revoked following a large number of public protests and a judicial decision, in 1940, which stated that he was morally unfit to teach at the College. Nine years later he was awarded the Order of Merit. He received the Nobel Prize for Literature in 1950.

Over a long and varied career, Bertrand Russell made ground-breaking contributions to the foundations of mathematics and to the development of contemporary formal logic, as well as to analytic philosophy. His contributions relating to mathematics include his discovery of Russell's paradox, his defence of logicism (the view that mathematics is, in some significant sense, reducible to formal logic), his introduction of the theory of types, and his refining and popularizing of the first-order predicate calculus.

Russell discovered the paradox which bears his name in May 1901, while working on his *Principles of Mathematics* (1903). The paradox arose in connection with the set of all sets which are not members of themselves.

Russell's own response to the paradox came with the introduction of his theory of types. His basic idea was that reference to troublesome sets (such as the set of all sets which are not members of themselves) could be avoided by arranging all sentences into a hierarchy.

Of equal significance during this same period was Russell's defence of logicism, the theory that mathematics was in some important sense reducible to logic. First defended in his *Principles*, and later in more detail in *Principia Mathematica*, Russell's logicism consisted of two main theses. The first is that all mathematical truths can be translated into logical truths or, in other words, that the vocabulary of mathematics constitutes a proper subset of that of logic. The second is that all mathematical proofs can be recast as logical proofs or, in other words, that the theorems of mathematics constitute a proper subset of those of logic.

During the 1950s and 1960s, Russell became something of an inspiration to large numbers of idealistic youth as a result of his continued anti-war and anti-nuclear protests. Together with Albert Einstein, he released the Russell–Einstein Manifesto in 1955, calling for the curtailment of nuclear weapons.

A2   The Axiom of Choice

The Axiom of Choice, first enunciated by Zermelo (1871–1953), is one of the most subtle of the axioms of set theory, and it has profound and mysterious implications. The books [JEC], [RR1], and [[RR2] treat the Axiom of Choice in great detail.

A2.1   Well ordering

It is not difficult to see that the well ordering of a set is closely related to the
Axiom of Choice (see Section 4.2 for terminology). For if S is a set that is well
ordered by a relation $\mathcal{R}$, then we may let $f : \mathcal{P}(S) \to S$ be the function that
assigns to each $A \in \mathcal{P}(S)$ the (perforce unique) minimal element of A. The
converse statement is true as well; but its proof involves ordinal arithmetic
and transfinite mathematical induction and cannot be treated at this time.

Here is a way to specify a well ordering of any countable set, such as the
rationals. The method seems artificial; but it *has* to be. The natural ordering
on these sets will not do the job.

Let S be a countable set. Let $\phi : S \to \mathbb{N}$ be a set-theoretic isomorphism.
If $x, y \in S$ are distinct elements, then we say that $(x, y) \in \mathcal{R}$ if $\phi(x) < \phi(y)$.
Check for yourself that this creates a strict, simple order on S that well orders
S.

As previously noted, it is impossible to give explicitly a well ordering of
the real numbers; however, it is a theorem that *any* set can be well ordered
(see [SUP]). The proof uses the Axiom of Choice.

A2.2   The continuum hypothesis

In Section 4.5 we briefly mentioned the continuum hypothesis: that there
are no cardinalities strictly between the cardinality of the integers and the
cardinality of the continuum (the cardinality of the reals). Call this statement
C. In 1938, Gödel showed that C could be added to the axioms of set theory
and no contradiction would ensue. That is to say, there is a "model" for set
theory in which the usual axioms of set theory are true and so is C. In 1963,
Paul Cohen showed that instead $\sim C$ could be added to the axioms of set
theory and no contradiction would ensue. That is, there is a model for set
theory in which the usual axioms of set theory are true but C is false (i.e.,
$\sim C$ is true).

In logical terms, we say that the continuum hypothesis is *independent* of
the other axioms of set theory. In particular, it is independent of the Axiom
of Choice. The assertion C can never be proved as a theorem from the other
axioms, nor can $\sim C$ be proved. Georg Cantor's inability to resolve the truth or
falsity of C was a strong contributing factor to his debilitating mental illness at
the end of his life. Sadly, mathematical logic was not sufficiently developed in
Cantor's time for him to have been able to understand the ultimate resolution
of the problem.

There are a number of very standard and useful tools in mathematics that
are consequences of the Axiom of Choice. We now enunciate two of them.

A2.3 Zorn's lemma

In modern mathematics, especially in algebra, Zorn's lemma plays a central role. It is used to prove the existence of maximal ideals, of bases for vector spaces, and of other "maximal sets."

We need two pieces of terminology in order to formulate Zorn's lemma. First, if $(S, \leq)$ is a partially ordered set, then a *chain* in S is a subset $C \subset S$ that is linearly ordered (i.e., any two elements are comparable). An element u is an upper bound of the chain C if $c \leq u$ for every $c \in C$.

A typical enunciation of Zorn's lemma is this:

Let $(S, \leq)$ be a non-empty, partially ordered set with the property that every chain in S has an upper bound. Then S has a maximal element, i.e., an element x such that $s \leq x$ for every $s \in S$.

Zorn's lemma is equivalent to the Axiom of Choice.

Zorn's lemma is commonly applied in algebra to prove the existence of various objects. For example, it can be used to establish the existence of a basis for any vector space (see [HER] for a discussion of these ideas, and for the associated terminology).

In what follows, a vector space V is a set with notions of addition and scalar multiplication. A set $\mathcal{B}$ is a *basis* for V if any element $v \in V$ can be written as $v = a_1 b_1 + a_2 b_2 + \cdots + a_k b_k$ for scalars a_j and elements $b_j \in \mathcal{B}$ and also the set $\mathcal{B}$ is minimal in size.

Proposition A2.1 *Let V be a vector space over the field F. Then V has a basis.*

Proof: Let $\mathcal{S}$ be the collection of all linearly independent sets in V. Partially order $\mathcal{S}$ as follows: if $A, B \in \mathcal{S}$, then $A \leq B$ if $A \subset B$. Now each chain clearly has a maximal element (or upper bound) since we may simply take the union of all the elements of the chain to be the upper bound. We then apply Zorn's lemma to conclude that the entire collection $\mathcal{S}$ has a maximal element. Call that maximal element Φ. We claim that Φ is a basis for V.

If the claim is not true, then there is an element x in V that is not in the span of Φ. But then x may be added to Φ, thereby contradicting the maximality of Φ. We conclude that Φ is the basis we seek. $\square$

Likewise, Zorn's lemma is used to establish the existence of maximal ideals in a ring, of algebraic closures, and of many other basic algebraic constructs. You will learn about these ideas in your course on abstract algebra.

A2.4 The Hausdorff maximality principle

The Hausdorff maximality principle is a variant of Zorn's lemma, also commonly used in algebraic applications.

Hausdorff's maximality principle: If $\mathcal{R}$ is a transitive relation on a set S then there exists a maximal subset of S which is linearly ordered by $\mathcal{R}$.

Hausdorff's principle is equivalent to the Axiom of Choice.

A2.5   The Banach–Tarski paradox

Just for fun, we conclude this section with a description of one of the most dramatic paradoxes in mathematics. This paradox stems from the Axiom of Choice, and is called the *Banach–Tarski paradox*. A version of this paradox is as follows:

Banach–Tarski paradox [HJE], [JEC]: It is possible to partition the solid ball in $\mathbb{R}^3$, of diameter one inch, into seven (disjoint) pieces in such a way that these seven pieces may be reassembled into a life-sized replica of the Statue of Liberty.

We refer the reader to the Bibliography (in particular [JEC]) for a detailed discussion of this paradox. We should note that the paradox fails in dimension two; it holds in dimension three in part because of the complexity of (that is, the non-abelian nature of) the three dimensional rotation group (as contrasted with the rather simple two dimensional rotation group).

Of course the seven pieces into which we break the unit ball in the Banach–Tarski paradox are extremely pathological. The subject of measure theory was invented, in part, to rule out sets such as these. Measure theory is another subject, like axiomatic set theory, in which there are very specific rules limiting the ways in which sets may be created.

A3   Independence and Consistency

We now turn to a brief and informal discussion of independence and consistency of the axioms of set theory. It is part of the spirit of the axiomatic method to have *as few axioms as possible*. For instance, it would be silly to have axioms that looked like this

1. $A \Rightarrow B$

2. $B \Rightarrow C$

3. $A \Rightarrow C$

because Axiom 3 follows logically from the other two axioms. It does not need to be enunciated as an axiom (although, conceivably, it might be useful later on to record it as a proposition). In this circumstance, we say that Axiom 3 is *not independent* of the other axioms.

Now the example that we have just presented is misleadingly simple. On the face of it, the axioms of set theory that we presented in the the first section of this Appendix all say something different. But it is conceivable, is it not, that after many pages of argument one might show that the first, third, and fifth axioms imply the sixth? How could we establish that such an eventuality cannot occur?

The question posed in the last paragraph was not considered formally until the twentieth century. The principal method that has evolved for showing that statement X is independent of axioms $A_1, A_2, \ldots, A_k$ is as follows.

1. One constructs a mathematical entity (such as a number system, or a geometry, or a version of set theory) in which $A_1, A_2, \ldots, A_k$ are true and X is also true.

2. One constructs a mathematical entity (such as a number system, or a geometry, or a version of set theory) in which $A_1, A_2, \ldots, A_k$ are true but X is false.

A moment's thought shows that $\sim X$ could not be proved as a theorem in the axiom system specified by $A_1, \ldots, A_k$, since (Statement **1**) we have constructed an instance of the axioms in which X is true (i.e., $\sim X$ fails). Likewise, statement X could not be proved as a theorem in the axiom system specified by $A_1 \ldots, A_k$ since (Statement **2**) we have constructed an instance of the axioms in which X fails (i.e., $\sim X$ is true).

All this may sound rather abstract, so let us consider a specific example. Euclidean geometry has five axioms. The first four are these:

P1. Through any pair of distinct points there passes a line.

P2. For each segment $\overline{AB}$ and each segment $\overline{CD}$ there is a unique point E (on the line determined by A and B) such that B is between A and E and the segment $\overline{CD}$ is congruent to $\overline{BE}$ (Figure A3.1).

P3. For each point C and each point A distinct from C there exists a circle with center C and radius CA (Figure A3.2).

P4. All right angles are congruent (Figure A3.3).

These are the standard four axioms that give our Euclidean conception of geometry. The fifth axiom, a topic of intense study for two thousand years, is the so-called parallel postulate (in Playfair's formulation):

P5. For each line ℓ and each point P that does not lie on ℓ there is a unique line ℓ' through P such that ℓ' is parallel to ℓ (Figure A3.4).

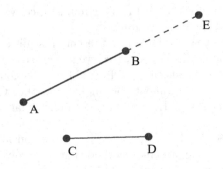

FIGURE A3.1
Euclid's second postulate.

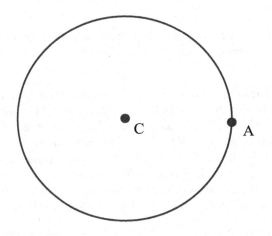

FIGURE A3.2
Euclid's third postulate.

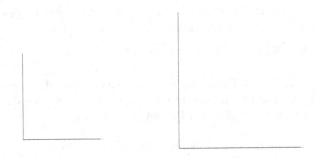

FIGURE A3.3
Euclid's fourth postulate.

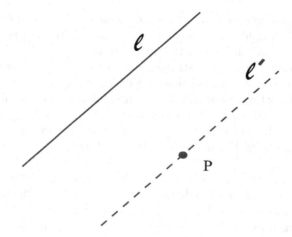

FIGURE A3.4
The parallel, or fifth, postulate of Euclid.

The "parallel postulate" of Euclidean geometry specifies that if ℓ is a line in the plane and P is a point not on that line, then there is one and only one line ℓ' passing through P that is parallel to ℓ. Classically, vigorous attempts were made to prove the parallel postulate as a *theorem* from the first four basic axioms of geometry. However, in the early nineteenth century, Nikolai Ivanovich Lobachevski (1793–1856) and János Bolyai (1775–1856) constructed a geometry in which the first four of the axioms of Euclid are satisfied yet the parallel postulate *fails* (see Section 9.2). Our standard (Euclidean/Cartesian) geometry is one in which all the axioms of Euclid are satisfied and the parallel postulate *holds*. The conclusion from the last two sentences is that the parallel postulate is *independent* of the other axioms of geometry. It cannot be proved as a theorem, nor can its negation be proved. What is true is that *either* the parallel postulate P *or* its negation $\sim P$ can be adjoined to the other axioms to form a valid geometry. More was said about these different geometries in Chapter 9.

Besides independence, the other big issue in setting up an axiomatic system is *consistency*. One could write down several plausible looking axioms which seem logical and start doing mathematics based on those axioms. Then one could wake up one day two years from now only to prove that $1 = 2$ or that $5 < 5$. This would be catastrophic.

How can we guarantee that an eventuality such as that described in the last paragraph can never happen? The answer comes by way of model theory. Think about how real life works. Two hundred or more years ago, the best scientists of the day gave cogent arguments that there could never be a heavier-than-air machine that flies. Today there are airplanes made of metal, that weigh many tons, and that fly. No matter how good the two-hundred-year-old (strictly philosophical) arguments are, and no matter how persuasive, the existence of metal airplanes lays the matter to rest.

Now let us get a bit more mathematical. Suppose I say to you that there is a number system containing the reals in which every polynomial equation has a root. When I say this, you should keep in mind that the polynomial $x^2+1 = 0$ has no *real roots*. We could debate this matter, from a philosophical point of view, at length. But in Chapter 5 we *constructed* a number system, called the complex numbers, in which this polynomial, indeed any polynomial, has a root.

The important word in the last paragraph is "construct." There is no value in saying "Yes, there is this number system that I dreamed up, and it contains a square root for -1 and other mysterious artifacts." We must construct the number system from mathematical tools at hand. This whole circle of ideas was treated in greater detail in Chapter 5.

Nikolai Lobachevski (1793–1856)

Nikolai Ivanovich Lobachevski was one of three sons in a poor family. In 1807, Lobachevski graduated from the Gymnasium and entered Kazan University. Lobachevski was highly successful in all courses he took. Martin Bartels (1769–1833) was Professor of Mathematics. Bartels was a school teacher and friend of Gauss. Gauss may have given Lobachevski mathematical hints through the letters between Bartels and Gauss. Lobachevski received a Master's Degree in physics and mathematics in 1811. In 1814 he was appointed to a lectureship and in 1816 he became an extraordinary professor.

In 1822 he was appointed as a full professor in the same year in which he began an administrative career as a member of the committee formed to supervise the construction of new university buildings. He was appointed to important positions within the university such as the Dean of the Mathematics and Physics Department between 1820 and 1825 and head librarian from 1825 to 1835. He also served as Head of the Observatory.

In 1827, Lobachevski became rector of Kazan University, a post he was to hold for the next 19 years. Two natural disasters struck the university while he was Rector of Kazan—a cholera epidemic in 1830 and a big fire in 1842. Owing to resolute and reasonable measures taken by Lobachevski, the damage to the University was reduced to a minimum. Lobachevski continued to teach a variety of different topics such as mechanics, hydrodynamics, integration, differential equations, the calculus of variations, and mathematical physics.

After Lobachevski retired in 1846, his health rapidly deteriorated. The illnes became progressively worse and led to blindness. These and financial difficulties added to the heavy burdens he had to bear over his last years. His great mathematical achievements were not recognised in his lifetime.

Lobachevski's major work, *Geometriya*, completed in 1823, was not published in its original form until 1909. On 11 February 1826, in the session of the Department of Physico-Mathematical Sciences at Kazan University, Lobachevski requested that his work about a new geometry be heard.

In 1837 Lobachevski published his article "Géométrie imaginaire" and a summary of his new geometry *Geometrische Untersuchungen zur Theorie der Parellellinien* was published in Berlin in 1840. This last publication greatly impressed Gauss.

There were two further major contributions to Lobachevski's geometry by Poincaré in 1882 and 1887. Perhaps these finally mark the acceptance of Lobachevski's ideas which would eventually be seen as vital steps in freeing the thinking of mathematicians so that relativity theory had a natural mathematical foundation.

This idea of constructing a *model* is the key to verifying consistency. When you write down a collection of axioms, you are describing a mathematical system or object that may or may not exist. But if you can provide an *independent construction* of the alleged object, one that satisfies the axioms, then you have explicitly exhibited an entity that fits the specified description. The axioms cannot lead to a contradiction because the model you have constructed shows that there are objects that possess the specified properties. Indeed, one of Kurt Gödel's fundamental theorems guarantees that any axiomatic system for which there is a model will never lead to a contradiction. [Again, this is a slight oversimplification; for all mathematical notions of consistency are *relative notions*. You will learn more about this idea in a course on formal logic.]

Model theory is an advanced and subtle idea. After you take an advanced course in logic, you may wish to refer to [CHK] for more on this topic.

A4 Set Theory and Arithmetic

We have already seen that the operation of set-theoretic product is analogous to multiplication in ordinary arithmetic. For the cardinality of $S \times T$ is the product of the cardinalities of S and T when the latter two are finite.

The other arithmetic operations may be modelled by set-theoretic operations. In this section we treat a few of these.

Definition A4.1 Let S and T be sets. Define T^S to be the set of all functions from S to T.

Example A4.2 Let $S = \{a, b, c\}$ and $T = \{1, 2\}$. The members of T^S are

$$\begin{aligned}
f_1 &= \{(a, 1), (b, 1), (c, 1)\} \\
f_2 &= \{(a, 1), (b, 1), (c, 2)\} \\
f_3 &= \{(a, 1), (b, 2), (c, 1)\} \\
f_4 &= \{(a, 1), (b, 2), (c, 2)\} \\
f_5 &= \{(a, 2), (b, 1), (c, 1)\} \\
f_6 &= \{(a, 2), (b, 1), (c, 2)\} \\
f_7 &= \{(a, 2), (b, 2), (c, 1)\} \\
f_8 &= \{(a, 2), (b, 2), (c, 2)\}
\end{aligned}$$

□

Notice in the last example that there are $8 = 2^3$ elements of T^S, which is precisely equal to $(\text{card } T)^{\text{card } S}$. This is an instance of a general phenomenon:

Proposition A4.3 *Let S and T each be sets with finite cardinality. Then*

$$\text{card}(T^S) = (\text{card } T)^{\text{card } S}.$$

Proof: Let the cardinality of T be k and the cardinality of S be m. Write

$T = \{t_1, \ldots, t_k\}$ and $S = \{s_1, \ldots, s_m\}$. If f is a function from S to T, then there are k possible values that $f(s_1)$ may take. Likewise, there are k possible values that $f(s_2)$ may take. Repeating this analysis a total of m times, we find that there are

$$\underbrace{k \cdot k \cdots k}_{m \text{ times}} = k^m$$

possible functions from S to T. □

Let us briefly treat the operation of set-theoretic product. If S and T are sets, then we have learned that $S \times T$ is the set of ordered pairs (s, t) such that $s \in S$ and $t \in T$. Once we note that $(S \times T) \times U$ can be identified with $S \times (T \times U)$ in a natural way, then we may meaningfully consider a *finite* product $S_1 \times S_2 \times \cdots \times S_k$. This is defined to be the set of all ordered k-tuples $(s_1, s_2, \ldots, s_k)$ such that $s_j \in S_j$ for $j = 1, \ldots, k$. As a simple example, Euclidean space $\mathbb{R}^3$ is nothing other than $\mathbb{R} \times \mathbb{R} \times \mathbb{R}$.

But now consider sets S_α for α in some possibly large (even infinite) index set A. We wish to define

$$\prod_{\alpha \in A} S_\alpha.$$

If A is an uncountable set, then it is essentially impossible to speak of ordered-tuples with the same cardinality as A. Yet it is frequently useful in mathematics to be able to treat this notion.

We motivate our new definition by recasting the product of two sets in a new language. Let S_1, S_2 be sets. Let us define $S_1 \times S_2$ to be the set of all functions $f : \{1, 2\} \to S_1 \cup S_2$ such that $f(1) \in S_1$ and $f(2) \in S_2$. Notice that each such function f can be thought of as the ordered pair $(f(1), f(2))$, with the first entry in S_1 and the second entry in S_2. So the language of functions allows us to express ordered pairs in a new way. Likewise, if T_1, T_2, T_3 are sets then we may define $T_1 \times T_2 \times T_3$ to be the set of all functions $f : \{1, 2, 3\} \to S_1 \cup S_2 \cup S_3$ such that $f(j) \in S_j, j = 1, 2, 3$. A moment's thought shows that this new definition is consistent with our notion of the product of three sets as discussed above. We may construct a similar definition for the product of any k sets, k a positive integer.

Now return to the consideration of a collection $\{S_\alpha\}_{\alpha \in A}$ of (possibly uncountably many) sets, indexed over a set A. Then the *product* of these sets is defined to be the set of all functions $f : A \to \left[\cup_{\alpha \in A} S_\alpha \right]$ such that $f(\alpha) \in S_\alpha$. We are no longer at liberty to compare this new definition with some more concrete set of ideas. For a large index set A, this must stand as our definition of product.

Example A4.4 For each $\alpha \in A = \mathbb{R}$, let $S_\alpha = \mathbb{Z}$. Then $\prod_{\alpha \in A} S_\alpha$ is just the collection of all functions from $\mathbb{R}$ into $\mathbb{Z}$ or, in the language introduced at the beginning of this section, $\mathbb{Z}^{\mathbb{R}}$. Of course this set has cardinality strictly greater than the cardinality of the continuum.

Exercises

1. Let S, T, U be finite sets. Prove that

$$\text{card}\left[\left[S^T\right]^U\right] = \text{card}\left[S^{T \times U}\right].$$

2. Let S, T, U be finite sets. Assume that $T \cap U = \emptyset$. Prove that

$$\text{card}\left[S^T \times S^U\right] = \text{card}\left[S^{T \cup U}\right].$$

* 3. What does our definition of the product of sets $\{S_\alpha\}_{\alpha \in A}$ have to do with the Axiom of Choice? Discuss.

4. Give examples of sets S, T, U, V such that $S \in T$, $T \in U$, $U \in V$. Use the ideas from Section A1 for inspiration.

* 5. Explain why the axioms of set theory, as presented in Section A1, disallow the construction of the set in Russell's paradox.

6. Give an explicit well ordering of the rational numbers. Do not use the result of Section A2 about the total ordering of countable sets to answer this question. [**Hint:** Refer to Exercises 4.11, 4.12 for the notion of lexicographic ordering.]

7. Let S and T be well ordered sets. Specify a well ordering of the set $S \times T$.

8. A set can be a subset of itself, but it cannot be an element of itself. Explain the difference, and why this situation is logically acceptable.

9. Prove that there is no infinite set that has cardinality smaller than $\mathbb{N}$. Do *not* use Proposition 4.5.24.

10. Suppose that we were to add an axiom to the set theory described in Section A1 that said "If S and T are sets, then so is $S \cap T$". This would be redundant (that is, it would not be independent of the other axioms). Explain why.

11. Suppose that we were to add an axiom to the set theory described in Section A1 that said "If S and T are sets, then so is $S \setminus T$". This would be redundant (that is, it would not be independent of the other axioms). Explain why.

12. Suppose that we were to add an axiom to the set theory described in Section A1 that said "If S and T are sets, then so is $S \times T$". This would be redundant (that is, it would not be independent of the other axioms). Explain why.

13. In some developments of set theory, there is an Axiom of Pairing that is formulated as follows: let x and y be objects (either sets or set elements); then

$$\exists A \; \forall z, z \in A \Leftrightarrow (z = x \vee z = y).$$

Prove that the Axiom of Pairing is a corollary of the Power Set Axiom and the Axiom Schema of Replacement.

* **14.** In some developments of set theory, there is an Axiom Schema of Separation that is formulated as follows: Let $P(x)$ be a property of x. Let z be a set. Then

$$\exists y \; \forall x, x \in y \Leftrightarrow \big(x \in z \wedge P(x)\big).$$

This axiom is important, for it specifies that a new set can only be defined as a subset of a known set. Prove that the Axiom Schema of Separation is a corollary of the Axiom Schema of Replacement.

* **15.** In some developments of set theory, there is a Union Axiom that is formulated as follows: Let A and B be sets; then

$$\exists C \; \forall x, x \in C \Leftrightarrow \big(x \in A \vee x \in B\big).$$

Prove that the Union Axiom is a corollary of the Axiom of Extensionality, the Axiom of Pairing (Exercise 13), and the Sum Axiom.

16. If A is a set and $A \subset A \times A$, then prove that $A = \emptyset$.

17. The mathematician/philosopher Bertrand Russell liked to say that if you need to choose one sock from each of infinitely many pairs of socks, then you will need to use the Axiom of Choice. But for shoes you do not. Explain what Russell meant by these statements.

18. Prove that if $A \times B = \emptyset$, then either $A = \emptyset$ or $B = \emptyset$.

* **19.** Let S and T be sets. Prove that not both $S \in T$ and $T \in S$.

20. If $S \times T = T \times S$, then what does that tell you about the sets S and T?

21. If A, B, C are sets and $A \subset B$, then prove that $A \times C \subset B \times C$.

22. If $A = \mathcal{P}(A)$, then what can you conclude about A?

23. Let X be a finite set. Explain why 2^X is an appropriate, and commonly used, notation for the power set of X.

24. A standard proof of the method of mathematical induction begins in this way. Suppose that $P(j)$ is a statement, that $P(1)$ is true, and that $P(j) \rightarrow P(j+1)$ for each j. Seeking a contradiction, assume now that it is *not the case* that $P(n)$ is true for every n. Let S be the set of all positive integers k such that $P(k)$ is false. Since $\mathbb{N}$ is well ordered, the set S has a least element. Complete the proof.

25. Explain why the methodology of Exercise 24 will not work to prove mathematical induction on the rationals or mathematical induction on the reals.

26. Give three distinct examples of partial orderings that are not total orderings.

27. Give three distinct examples of total orderings that are not well orderings.

28. Identify the type of each of the following orderings $\triangle$:

(a) S and T are sets; $S \triangle T$ if $S \supset T$.

(b) m and n are integers; $m \triangle n$ if $m + n > 5$.

(c) x and y are real numbers; $x \triangle y$ if $x \cdot y > 0$.

(d) S and T are sets; $S \triangle T$ if $S \in T$.

(e) z and w are real numbers; $z \triangle w$ if $z - w$ is a positive real number.

29. Let S be the collection of all sets that can be described in fewer than 50 words. Is S an element of itself? Why or why not? What does this example have to do with Russell's paradox? Do the axioms of set theory forbid this set?

Chapter 1: Basic Logic

1. (a)

S	T	S∧T	~(S∨T)	(S∧T)∨~(S∨T)
T	T	T	F	T
T	F	F	F	F
F	T	F	F	F
F	F	F	T	T

(c)

S	T	~S∨T	~(S∧~T)	(~S∨T)⇔~(S∧~T)
T	T	T	T	T
T	F	F	F	T
F	T	T	T	T
F	F	T	T	T

(e)

S	T	~S⇒T	S⇒(~S⇒T)	S⇒(~S⇒(S⇒ (~S⇒T)))
T	T	T	T	T
T	F	T	T	T
F	T	T	T	T
F	F	F	T	T

(g)

S	T	T∨~S	S∧(T∨~S)
T	T	T	T
T	F	F	F
F	T	T	F
F	F	T	F

2. (c) $(\sim V \vee \sim T) \Rightarrow \sim U$

(e) $S \Rightarrow \sim T$

(g) $\sim (\sim S \vee \sim U)$

(i) $(\sim V \wedge \sim U) \Rightarrow \sim S$

(k) $\sim T \Rightarrow \sim U$

3. (a) If all politicians are honest and no man is a fool, then I do have two brain cells to rub together.

   (c) If the pie is in the sky then, for all politicians to be honest, it is necessary that some men be fools.

   (e) The pie is in the sky and I don't have two brain cells to rub together if and only if the pie is in the sky or I don't have two brain cells to rub together.

   (g) Either the pie is in the sky or, if some politicians are not honest, then it must be that some men are fools.

   (i) I don't have two brain cells to rub together if and only if it is necessary and sufficient that the pie be in the sky for me not to have two brain cells to rub together.

5. Recall that the statement $S \Rightarrow T$ is true if S is false or if T is true and S is true as well.

   (a) True, since "π is rational" is false.

   (c) True, since "$2 + 2 = 5$" is false.

   (e) True, since "$4^2 = 16$" is true.

   (h) False, since "it is not the case that $3^2 = 8$" is true, but "$4^2 = 17$" is false.

   (i) False, since "both $3 \cdot 2 = 6$ and $4+4 = 8$" is true, but "$5 \cdot 5 = 20$" is false.

6. One only has to substitute $\sim (\sim S \vee \sim T)$ for $S \wedge T$ (De Morgan's Law) and $\sim S \vee T$ for $S \Rightarrow T$.

   (a) $\sim S \vee \sim T$.

   (b) $\sim (S \vee T)$.

   (e) $\sim (S \vee T) \vee \sim (\sim S \vee \sim T)$.

   (g) $(\sim S \vee T) \vee (\sim T \vee S)$.

8. If S is false and T is false, then $S \Rightarrow T$ is true while $S \wedge T$ is false and $S \vee T$ is false. So it cannot be done.

9. (a) $\exists x, x > 0 \wedge x^2 = 5$

   (c) $\sim (\exists x \in \mathbb{N} \exists y \in \mathbb{N} \exists m \in \mathbb{N} \exists z \in \mathbb{N}, m = x^3 + y^3 \wedge m = z^3)$

   (f) $z \in \mathbb{C} \wedge w \in \mathbb{C} \Rightarrow z \cdot w \in \mathbb{C}$

   (h) $\sim ((x \in \mathbb{R} \setminus \mathbb{Q} \wedge y \in \mathbb{R} \setminus \mathbb{Q}) \Rightarrow (x \cdot y \in \mathbb{R} \setminus \mathbb{Q}))$

   (j) $x \in \mathbb{Q} \Rightarrow x^2 \in \mathbb{Q}$

10. (a) There is a positive real that has no positive square root.

    (c) There are two numbers such that the sum of their squares is strictly less than twice their product.

(e) There is a real number greater than 1 that is greater than its square.

(g) There is a number such that one plus this number has no cube root.

11. $\exists m_1 \exists m_2 \exists m_3 \exists m_4 \exists m_5, (P(m_1) = 0 \wedge P(m_2) = 0 \wedge P(m_3) = 0 \wedge P(m_4) = 0 \wedge P(m_5) = 0)$.

13. (a) The set S contains at most one integer.

(c) Either I am not rough or I am not tough or I do not breathe fire.

(e) Either I will not marry Fred or I will please Irving.

(g) Either I will not pay my taxes or I will go to jail.

(i) Either I do not love everyone or not everyone loves me.

(k) If you get caught then you will not go to jail.

(m) If you make more than \$ 100,000, then you will pay income tax.

14. Only (d) and (f) are logically equivalent.

17. The statement $A \wedge B$ is logically equivalent with $\sim (\sim A \vee \sim B)$. The statement $A \Rightarrow B$ is logically equivalent with $\sim A \vee B$. So any statement involving $\wedge, \Rightarrow, \vee, \sim$ is logically equivalent to a statement containing only the connectives $\sim$ and $\vee$.

18. The statement $A \vee B$ is logically equivalent with $\sim A \Rightarrow B$. The statement $A \wedge B$ is logically equivalent with $\sim (A \Rightarrow \sim B)$. So any statement involving $\wedge, \Rightarrow, \vee, \sim$ is logically equivalent to a statement containing only the connectives $\sim$ and $\Rightarrow$.

19. Suppose that statement S is true. Then $\sim S$ is false. But $S \vee S$ and $S \wedge S$ are both true. So there is no way to obtain a statement logically equivalent to $\sim S$ using $\vee$ and $\wedge$.

20. One example is the connective "nor," which is defined to be

$$ S \downarrow T \equiv \sim (S \vee T). $$

Check that $S \downarrow S$ is logically equivalent to $\sim S$ and $[S \downarrow S] \downarrow [T \downarrow T]$ is logically equivalent to $S \wedge T$. We know from Exercise 1.16 that $\sim$ and $\wedge$ are sufficient to formulate all statements in sentential logic.

22. The statement $\forall x, P(x)$ says that P is true for every value of the argument x. The statement $\sim \exists \sim P(x)$ says that there is no value of x for which $P(x)$ fails. These say the same thing.

23. The statement $\exists x, Q(x)$ says that there is some value of x for which $P(x)$ is true. The statement $\sim \forall \sim Q(x)$ says that it is not true that $Q(x)$ fails for all x. These say the same thing.

24. The statement $\forall x \exists y, y > x$ says that, for any x, there is a y that is greater than x. In the real numbers, for instance, this is true.

The statement $\exists y \forall x, y > x$ says that there is a y that is greater than all x. In the real numbers, for instance, this is false. So the two statements are quite different, and we cannot commute the quantifiers.

Chapter 2: Methods of Proof

1. If m, n are odd then, by definition, $m = 2k + 1$ and $n = 2\ell + 1$ for some $k, \ell \in \mathbb{Z}$. Multiplying, we obtain:

$$m \cdot n = (2k + 1) \cdot (2\ell + 1) = 2(2k\ell + k + \ell) + 1,$$

which is odd by definition.

2. If n is even then, by definition, $n = 2\ell$ for some $\ell \in \mathbb{Z}$. Multiplying, we obtain:

$$m \cdot n = m(2\ell) = 2(m\ell),$$

which is even by definition.

4. Writing things out, we obtain

$$
\begin{aligned}
2 + 4 + \cdots + (2k - 2) + 2k &= 2\left(1 + 2 + \cdots + (k - 1) + k\right) \\
&= 2\,\frac{k(k + 1)}{2} \qquad \text{by Prop. 2.4.2} \\
&= k^2 + k.
\end{aligned}
$$

5. Writing things out, we obtain

$$
\begin{aligned}
1 + 3 + \cdots &+ (2k - 3) + (2k - 1) \\
&= (2 - 1) + (4 - 1) + \cdots + ((2k - 2) - 1) + (2k - 1) \\
&= (2 + 4 \cdots + (2k - 2) + 2k) - k \\
&= (k^2 + k) - k \qquad\qquad\qquad \text{by Exercise 2.4} \\
&= k^2.
\end{aligned}
$$

7. We can write $m = 3^k$ and $n = 3^\ell$ for some $k, \ell \in \mathbb{N}$. Let us assume, without loss of generality, that $k \le \ell$. Seeking a contradiction, assume that $m + n = 3^r$, for some $r \in \mathbb{N}$. Then we have:

$$m + n = 3^\ell + 3^k = 3^k(1 + 3^{\ell-k}) = 3^r, \quad \text{with} \quad \ell - k \ge 0.$$

Now note that since $3^k = m < m + n = 3^r$, it must be that $k < r$. Hence,

$$1 + 3^{\ell-k} = 3^{r-k}, \quad \text{with} \quad r - k \ge 0.$$

The right-hand side of the last equality is either 1 or divisible by 3, whereas the left-hand side is bigger than or equal to 2 and definitely not divisible by three. Thus, the equality must be false, which means that our original hypothesis $m + n = 3^r$, for some $r \in \mathbb{N}$) was false.

11. Following the given scheme, we have $2q^2 = p^2$. If q has, say, r prime factors, then q^2 has $2r$ prime factors. Thus $2q^2$ has $2r + 1$ prime factors. On the other hand, p^2 must have an even number of prime factors, and we arrive at a contradiction. Hence our original assumption $\sqrt{2} = p/q$ must be false.

12. Write $n = k^2$, and suppose that $n + 1 = \ell^2$ for some $l \in \mathbb{N}$. Then $1 = (n+1) - n = \ell^2 - k^2 = (\ell + k)(\ell - k)$. Then $\ell + k = 1$, and that is impossible for natural numbers ℓ and k.

13. We proved in Exercise 2.1 that the product of two odd numbers is odd. Therefore, if the product of two numbers is even, at least one of them must be even.

15. If n is even, then $n - 1$ is odd. Now, $n = (n - 1) + 1$, so n is the sum of two odds. If n is odd, then $n = n$, so n is the sum of one odd integer.

17. False: $1^2 + 2^2 = 5$, which is not a perfect square.

18. True: Suppose, seeking a contradiction, that there are no perfect squares in that list. This means that all the numbers in the list fall between two consecutive squares, i.e., there is a k such that $k^2 < n < n+1 < \ldots < 2n + 2 < (k + 1)^2$. We then have:

$$2n + 2 < (k + 1)^2 = k^2 + 2k + 1 < n + 2k + 1,$$

or

$$n + 1 < 2k.$$

Squaring both sides, we obtain:

$$n^2 + 2n + 1 < 4k^2 < 4n,$$

or

$$(n - 1)^2 = n^2 - 2n + 1 < 0,$$

which is impossible.

19. True: 6=1+2+3, or 28=1+2+4+7+14. In fact, these numbers have a name: Perfect numbers. Not much is known about perfect numbers. It is conjectured that there are no odd perfect numbers, mainly because nobody ever found an odd perfect number, but it has never been proved. It has also not been proved that there are infinitely many perfect numbers.

21. False: $2^2 + 1^2 = 5$, which is a prime.

23. False: Take $n = 2, a_1 = 1, a_2 = 4$. Then the inequality would read:

$$\frac{5}{2} = \frac{1+4}{2} \le (1 \cdot 4)^{1/2} = 2,$$

which is clearly false.

25. True: Write the rationals as fractions p_1/q, p_2/q with the same denominator q so that $p_1 + 2 \le p_2$ (one can always achieve this by taking q big enough). Then either $\sqrt{p_1^2 + 1}$ or $\sqrt{p_1^2 + 2}$ is an irrational number that lies between p_1 and p_2 (cf. Exercise 2.12). Divide this irrational by q to obtain another irrational that lies between the two rationals.

27. We will use the alternative form of the principle of complete mathematical induction given in the text (see also Exercise 2.31). The property is clearly true for 2 (since 2 is prime). Assume that it is true for any $k < n$. We have to prove that it is true for $k = n$. If the only divisors of n are n itself and 1, we are done, since that would imply that n is itself a prime. Otherwise n has a divisor $d < n$. By the mathematical induction hypothesis, d must have some prime factor p. Now, since p divides d which divides n, p must also divide n. But then p will be a factor of n. Hence the property is also true for n.

29. The property is true for $k = 3$, since $2^3 = 8 > 7 = 1 + 2 \cdot 3$. Assume that the property is true for $k = n - 1$. We want to show that it is true for $k = n$. In other words, we want to prove that

$$2^n > 1 + 2n.$$

Observing that $2^n = 2 \cdot 2^{n-1}$, we find an obvious place to apply the mathematical induction hypothesis:

$$\begin{aligned}
2^n &= 2 \cdot 2^{n-1} \\
&> 2 \cdot (1 + 2(n-1)) \\
&\quad \text{from the mathematical induction hypothesis} \\
&= (1 + 2n) + (2n - 3) \\
&> 1 + 2n \quad \text{since } 2n - 3 > 0 \text{ for } n \ge 2.
\end{aligned}$$

31. If one starts the mathematical induction process from a number $n_0 + 1$, $n_0 \ge 1$, then $P(1)$ might not be true (it might not even be defined), and we would not be able to use mathematical induction according to the statement in the text. But we can modify this statement using the following trick:

Define a property P' as follows:

$$P'(k) \text{ is true} \quad if \quad P(n_0 + k) \text{ is true}.$$

Then $P'(1)$ is true since $P(n_0 + 1)$ is true, and $P'(n-1) \Rightarrow P'(n)$ because $P(n_0 + n - 1) \Rightarrow P(n_0 + n)$ by hypothesis. But now we can apply mathematical induction (as stated in the text) to the property P', so that $P'(n)$ holds for any natural number n. This implies that $P(n_0 + n)$ holds for all $n \in \mathbb{N}$, or equivalently, $P(m)$ holds for all $n \geq n_0 + 1$.

34. It is true for $n = 5$:

$$2^5 = 32 > 26 = 5^2 + 1.$$

Assume that it is true for $n - 1 \geq 5$. We must prove it for n. We can write:

$$
\begin{aligned}
2^n &= 2 \cdot 2^{n-1} \\
&> 2((n-1)^2 + 1) \quad \text{by the mathematical induction hypothesis} \\
&= n^2 + 1 + (n^2 - 4n + 3) \\
&= n^2 + 1 + (n^2 - 4n + 4) - 1 \\
&= n^2 + 1 + (n-2)^2 - 1 \\
&> n^2 + 1 \quad \text{since } (n-2)^2 - 1 > 0 \text{ if } n > 4.
\end{aligned}
$$

35. Let $P(n)$ be the statement, "If $n + 1$ letters are placed into n mailboxes, then some mailbox must contain two letters. When $n = 1$, the claim is that if we put two letters into one mailbox, then some mailbox must contain two letters. Obvious. Now suppose that $P(n-1)$ has been proved. We have n mailboxes, and we place $n + 1$ letters into n mailboxes. If the last mailbox contains two letters, then we are done. If not, then the last box contains one or two letters. But then the first $n-1$ mailboxes contain at least n letters. So the inductive hypothesis applies, and one of them must contain two letters. That completes the inductive step, and the proof.

36. Assume that we have n mailboxes. Let $\ell(j)$ be the number of letters in box j. Now

$$\ell(1) + \ell(2) + \cdots + \ell(n) = n + 1,$$

since all the letters taken together total $n + 1$ letters. Dividing by n gives

$$\frac{\ell(1) + \ell(2) + \cdots + \ell(n)}{n} = \frac{n+1}{n} > 1.$$

So the average number of letters per box exceeds 1. This can only be true if some box contains more than 1 letter. Thus some box contains two letters.

38. Consider the set of S all ordered pairs (ℓ, p) where ℓ is a line passing through (at least) two of the given points and p is a point not on that line (certainly p exists because the points are not all colinear). Define a function f on S by

$$f(\ell, p) = \text{distance of } \ell \text{ to } p.$$

Then f is a function with a finite domain, so there is a particular ordered pair (ℓ_0, p_0) that minimizes the function. Then ℓ_0 is the line that we seek. We invite the reader to check cases to verify this assertion.

40. Let

$$B(x) = x \text{ is a boy under the age of } 10.$$

and

$$P(x) = x \text{ practices all pieces in his/her piano book every day.}$$

Then our statement is

$$\forall x, B(x) \Rightarrow P(x).$$

We can rewrite this as

$$\sim \exists x, \sim (B(x) \Rightarrow P(x)).$$

41. The assertion is true for $n = 1$ by inspection.

Assume now that the assertion is verified for $n = j$. Then we have

$$\begin{pmatrix} a & 2 \\ 0 & a \end{pmatrix}^{j+1} = \begin{pmatrix} a & 2 \\ 0 & a \end{pmatrix} \cdot \begin{pmatrix} a & 2 \\ 0 & a \end{pmatrix}^{j}$$

$$= \begin{pmatrix} a & 2 \\ 0 & a \end{pmatrix} \cdot \begin{pmatrix} a^j & 2ja^{j-1} \\ 0 & a^j \end{pmatrix}$$

$$= \begin{pmatrix} a^{j+1} & 2(j+1)a^j \\ 0 & a^{j+1} \end{pmatrix}.$$

That completes the inductive step.

42. The assertion is clear for $n = 1$. Now assume that it is true for $n = j$. We write

$$\begin{aligned} (j+1)^3 - (j+1) &= (j^3 + 3j^2 + 3j + 1) - (j+1) \\ &= j^3 + 3j^2 + 2j \\ &= (j^3 - j) + (3j^2 + 3j). \end{aligned}$$

Now, by the inductive hypothesis, $j^3 - j$ is divisible by 6. Also

$$3j^2 + 3j = 3j(j+1).$$

Since either j or $j + 1$ is divisible by 2, this last expression is also divisible by 6. Hence $(j + 1)^3 - (j + 1)$ is divisible by 6, and the mathematical induction is complete.

44. We will prove that, for any positive integer n,

$$\frac{1}{\sqrt{1}} + \frac{1}{\sqrt{2}} + \cdots + \frac{1}{\sqrt{n}} \geq \sqrt{n}.$$

The claim is plainly true for $n = 1$. Now assume that it has been established for $n = j$. Then we have

$$\frac{1}{\sqrt{1}} + \frac{1}{\sqrt{2}} + \cdots + \frac{1}{\sqrt{j}} + \frac{1}{\sqrt{j+1}} \geq \sqrt{j} + \frac{1}{\sqrt{j+1}}.$$

So we need to show that

$$\sqrt{j} + \frac{1}{\sqrt{j+1}} \geq \sqrt{j+1}.$$

Multiplying both sides by $\sqrt{j+1}$, we see that this is the same as

$$\sqrt{j(j+1)} + 1 \geq j + 1$$

or

$$\sqrt{j(j+1)} \geq j.$$

Now squaring both sides gives the result. The mathematical induction is complete, and the result proved.

Chapter 3: Set Theory

1. (a) $S \cap U = \{1, 2, 3, 4\}$.

(b) $(S \cup U) \cap V = \{2, 4\}$.

(c) $(U \cup V \cup T) \setminus S = \{6, 7, 8, 9\}$.

(d) $(S \times V) \setminus (T \times U) = \{(1, 2), (1, 4), (1, 6), (1, 8), (2, 2), (2, 4),$
$(2, 6), (2, 8), (3, 6), (3, 8), (4, 6), (4, 8), (5, 6), (5, 8)\}$.

4. $\subseteq$:If $x \in {}^c(\cap_{\alpha \in A} S_\alpha)$, then $x \notin \cap_{\alpha \in A} S_\alpha$, which implies $\exists \alpha_0$ such that $x \notin S_{\alpha_0}$. But this means that $x \in {}^c S_{\alpha_0}$, so certainly $x \in \cup_{\alpha \in A} {}^c S_\alpha$.

$\supseteq$:If $x \in \cup_{\alpha \in A} {}^c S_\alpha$, then $\exists \alpha_0$ such that $x \in {}^c S_{\alpha_0}$. Thus, $x \notin S_{\alpha_0}$, so $x \notin \cap_{\alpha \in A} S_\alpha$. Therefore, $x \in {}^c(\cap_{\alpha \in A} S_\alpha)$.

$\subseteq$:If $x \in T \cap (\cup_{\alpha \in A} S_\alpha)$, then $x \in T$ and $x \in \cup_{\alpha \in A} S_\alpha$. This implies that $x \in T$ and $x \in S_{\alpha_0}$ for some $\alpha_0 \in A$. So $x \in T \cap S_{\alpha_0}$, which implies $x \in \cup_{\alpha \in A} (T \cap S_\alpha)$.

$\supseteq$:If $x \in \cup_{\alpha \in A}(T \cap S_\alpha)$, then $x \in T \cap S_{\alpha_0}$ for some $\alpha_0 \in A$. Thus, $x \in T$ and $x \in S_{\alpha_0}$. Therefore, $x \in T$ and $x \in \cup_{\alpha \in A} S_\alpha$, so $x \in T \cap (\cup_{\alpha \in A} S_\alpha)$.

5. $S \cap (T \cup U) = (S \cap T) \cup (S \cap U)$:

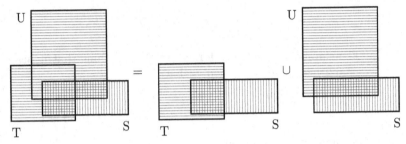

$S \cap {}^c T = S \setminus T$

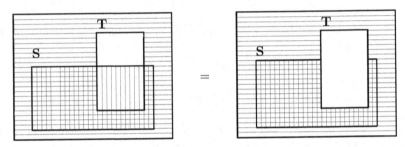

7. $\mathbb{Q} \setminus \mathbb{Z} = \{p/q \in \mathbb{Q}$ such that q does not divide $p\}$.

$\mathbb{R} \setminus \mathbb{Q} = \{x \in \mathbb{R}$ such that $x \neq p/q$ for any $p, q \in \mathbb{Z}\}$. These are simply the irrational real numbers.

9. $(\mathbb{Q} \times \mathbb{R}) \setminus (\mathbb{Z} \times \mathbb{Q}) = \{(\mathbf{x}, \mathbf{y})$ such that $\mathbf{x} \in \mathbb{Q}$ and $\mathbf{y} \in \mathbb{R}$, but either $x \notin \mathbb{Z}$ or $y \notin \mathbb{Q}\}$.

12. $\mathcal{P}(T) = \{\emptyset, \{1\}, \{2\}, \{1, 2\}\}$. Therefore

$$\mathcal{P}(\mathcal{P}(T)) = \Big\{ \emptyset, \{\emptyset\}, \{\{1\}\}, \{\{2\}\}, \{\{1, 2\}\}, \{\emptyset, \{1\}\}, \{\emptyset, \{2\}\},$$
$$\{\emptyset, \{1, 2\}\}, \{\{1\}, \{2\}\}, \{\{1\}, \{1, 2\}\}, \{\{2\}, \{1, 2\}\},$$
$$\{\emptyset, \{1\}, \{2\}\}, \{\emptyset, \{1\}, \{1, 2\}\}, \{\emptyset, \{2\}, \{1, 2\}\},$$
$$\{\{1\}, \{2\}, \{1, 2\}\}, \{\emptyset, \{1\}, \{2\}, \{1, 2\}\} \Big\}.$$

13. False: Take $S_j = Set-(j-1)j-1$, for $j > 1$ and $S_1 = \{0\}$. All these sets have finitely many elements, yet their union is all of $\mathbb{Z}$.

15. $S \subset T \Rightarrow \mathcal{P}(S) \subset \mathcal{P}(T)$: Suppose that $R \in \mathcal{P}(S)$. Then, by definition of power set, we must have that $R \subset S$. But $S \subset T$, so by Exercise 3.17 below we must have that $R \subset T$, which implies $R \in \mathcal{P}(T)$.

$\mathcal{P}(S) \subset \mathcal{P}(T) \Rightarrow S \subset T$: Suppose that $x \in S$. Then $\{x\} \in \mathcal{P}(S) \subset \mathcal{P}(T)$. Thus $\{x\} \in \mathcal{P}(T)$, which implies that $x \in T$.

17. Suppose that $x \in A$. Then $x \in B$ since $A \subset B$. But then $x \in C$ since $B \subset C$. Therefore, $A \subset C$.

19. (a)

$$\mathcal{P} = \Big\{ \emptyset, \{1\}, \{\emptyset\}, \{\{a, b\}\}, \{1, \emptyset\}, \{1, \{a, b\}\}, \{\emptyset, \{a, b\}\},$$
$$\{1, \emptyset, \{a, b\}\} \Big\}$$

(b)

$$\mathcal{P} = \Big\{ \triangle, \{\bullet\}, \{\triangle\}, \{\partial\}, \{\bullet, \triangle\}, \{\bullet, \partial\}, \{\triangle, \partial\}, \{\bullet, \triangle, \partial\} \Big\}$$

(c)

$$\mathcal{P} = \Big\{ \{\emptyset\}, \{\emptyset\}, \{\{\emptyset\}\}, \{\{\{\emptyset\}\}\}, \{\emptyset, \{\emptyset\}\}, \{\emptyset, \{\{\emptyset\}\}\}, \{\{\emptyset\},$$
$$\{\{\emptyset\}\}\}, \{\emptyset, \{\emptyset\}, \{\{\emptyset\}\}\} \Big\}$$

21. (a) This is false in general. Let $A = \{1, 2\}$ and $B = \{3, 4\}$. Then $\{1, 4\} \in \mathcal{P}(A \cup B)$ but $\{1, 4\} \notin \mathcal{P}(A) \cup \mathcal{P}(B)$.

(c) This is false in general. Let $A = \{1, 2, 3\}$ and $B = \{1\}$. Then $A \setminus B$ will have 2 elements and $\mathcal{P}(A \setminus B)$ will therefore have 4 elements. On the other hand, $\mathcal{P}(A)$ will have 8 elements and $\mathcal{P}(B)$ will have 2 elements. And $4 \neq 8 - 2$.

23. Suppose that $x \in \cup_{\alpha \in A} T_\alpha$. Then $x \in T_{\alpha_0}$ for some $\alpha_0 \in A$. Now, by hypothesis, $T_{\alpha_0} \subset S$, so $x \in S$.

25. Suppose that S has j elements, T has k elements, and U has m elements. Then $S \times T$ has $j \cdot k$ elements and $(S \times T) \times U$ has $(j \cdot k) \cdot m$ elements. On the other hand, $T \times U$ has $k \cdot m$ elements and $S \times (T \times U)$ has $j \cdot (k \cdot m)$ elements. Of course multiplication of integers is associative, so $(j \cdot k) \cdot m = j \cdot (k \cdot m)$. As a result, the two indicated sets have the same number of elements.

27. Let $T = \{1\}$. Then $S \times T = \{(s, 1) : s \in S\}$, and $S \times T$ can easily be put in one-to-one correspondence with S itself. Since $S \times T$ is finite by hypothesis, then so is S.

29. The set $\mathcal{P}$ has $2^5 = 32$ subsets. There are 2^{32} subsets of $\mathcal{P}$. But only 32 of these can be power sets of subsets of I. So the requested probability is $32/2^{32} \approx 7.45 \times 10^{-9}$.

Chapter 4: Relations and Functions

1. Reflexive : $m + m = 2m$ is even, so $(m, m) \in \mathcal{R}$.

Symmetric : If $m + n$ is even, then $n + m$ is also even, so $m\mathcal{R}n \Rightarrow n\mathcal{R}m$.

Transitive : If $m + n$ and $n + p$ are both even, then either m, n, p are all even or they are all odd. In either case, $m + p$ will also be even. So $m\mathcal{R}n$ and $n\mathcal{R}p \Rightarrow m\mathcal{R}p$.

To find the equivalence classes, note that an even and an odd are never in the same class, whereas two odds or two evens are always in the same class. So there are two equivalence classes: The class of even numbers and the class of odd numbers.

4. Reflexive : $y = y$, so $(x, y)\mathcal{R}(x, y)$.

Symmetric : If $y = y'$, then $y' = y$, so $(x, y)\mathcal{R}(x', y') \Rightarrow (x', y')\mathcal{R}(x, y)$.

Transitive : If $y = y'$ and $y' = y''$, then $y = y''$, so $(x, y)\mathcal{R}(x', y')$ and $(x', y')\mathcal{R}(x'', y'') \Rightarrow (x, y)\mathcal{R}(x'', y'')$.

To each element (x, y) assign the representative $(0, y)$. This assignment takes the set of equivalent classes in a one-to-one and onto fashion into the set $\{0\} \times \mathbb{R}$. Another way to see what the equivalence classes are is to note that two members of the same class always lie in the same horizontal line. Thus the set of equivalence classes is a horizontal line.

5. Reflexive: $x - x = y - y = 0 \in \mathbb{Z}$, so $(x, y)\mathcal{R}(x, y)$.

Symmetric: If $x - x' \in \mathbb{Z}$ and $y - y' \in \mathbb{Z}$, then $x' - x = -(x - x') \in \mathbb{Z}$ and $y - y' = -(y - y') \in \mathbb{Z}$, so $(x, y)\mathcal{R}(x', y') \Rightarrow (x', y')\mathcal{R}(x, y)$.

Transitive: If $x - x' \in \mathbb{Z}$ and $y - y' \in \mathbb{Z}$, and $x' - x'' \in \mathbb{Z}$ and $y' - y'' \in \mathbb{Z}$, then $x - x'' = (x - x') + (x' - x'') \in \mathbb{Z}$ and $y - y'' = (y - y') + (y' - y'') \in \mathbb{Z}$, so $(x, y)\mathcal{R}(x', y')$ and $(x', y')\mathcal{R}(x'', y'') \Rightarrow (x, y)\mathcal{R}(x'', y'')$.

To see what equivalence classes are, note that $(x, y)\mathcal{R}(x + m, y + n)$, where $m, n \in \mathbb{Z}$. Thus, each equivalence class is an infinite array of dots, like the points of a grid (this is often called a *lattice*). By assigning to each class the representative with both coordinates non-negative and with least x and y coordinates, we find a one-to-one and onto correspondence between the set of equivalence classes and the square $[0,1) \times [0,1)$. The set of equivalence classes is actually isomorphic to a torus (the surface of a doughnut). (Thinking of how this could be true can be a fun exercise in geometric visualization.)

7. Two people are related by blood if they have a common ancestor.

Reflexive : Clear: One always has the same parents as oneself.

Symmetric : If person α has blood relative X in common with person β, then of course β has blood relative X in common with person α.

NOT Transitive : It is possible for person α to have a father F^1 in common with person β (father F^1 sired α with one mother M^1 and sired β with another mother M^2). Then person β can have mother M^2 in common with person γ (mother M^2 gave birth to β and γ with a different father, F^2). But α and γ will have no blood relative in common.

Thus the relation is not transitive.

8. Reflexive: If (m,n) is an ordered pair of rationals, then $m \cdot n = m \cdot n$. Hence $(m,n)\mathcal{R}(m,n)$.

Symmetric: If $(m,n)\mathcal{R}(m',n')$, then $m \cdot n' = m' \cdot n$, hence $m' \cdot n = m \cdot n'$. So $(m',n')\mathcal{R}(m,n)$.

Transitive If $(m,n)\mathcal{R}(m',n')$ and $(m',n')\mathcal{R}(m'',n'')$, then $m \cdot n' = m' \cdot n$ and $m' \cdot n'' = m'' \cdot n'$. Note that if $m = 0$, then the first equation implies that $m' = 0$ and the second equation then implies that $m'' = 0$. So all three ordered pairs are related. Thus we may assume that none of m, m', m'' is zero. Now let us multiply our two equations together: Left side against left side and right side agains right side. The result is $m \cdot n' \cdot m' \cdot n'' = m' \cdot n \cdot m'' \cdot n'$. Finally we may cancel m' and n' from both sides (since both these numbers are nonzero). So $m \cdot n'' = m'' \cdot n$ and we see that $(m,n)\mathcal{R}(m'',n'')$. The relation is transitive.

Every equivalence class has an element with second entry $n = 1$. So we may identify $[(m,1)] \leftrightarrow m$. Thus the equivalence classes identify naturally with the rational numbers. As a result, our equivalence relation does not create anything new.

9. All of them are functions except **(f)**, which is not a function because it assigns two different values (6 and 5) to c, and **(b)** and **(g)**, because they do not use all the elements of the domain.

13. It can be defined as follows:

Let $\alpha = (a_1, a_2, \ldots, a_k)$ and $\alpha' = (a'_1, a'_2, \ldots, a'_k)$ be distinct k-tuples. Let i_0 be the first integer such that $a_{i_0} \neq a'_{i_0}$ (such an i_0 should exist since α and α' are assumed to be distinct). Write $\alpha \leq \alpha'$ if $a_{i_0} \leq a'_{i_0}$ and $\alpha' \leq \alpha$ if $a'_{i_0} \leq a_{i_0}$. It can be easily checked that this gives a partial ordering.

15. (b) is expressed as:

$$f - g = \{(s, t - t') : (s, t) \in f, \ (s, t') \in g\}.$$

(c) is expressed as:

$$f \cdot g = \{(s, t \cdot t') : (s, t) \in f, \ (s, t') \in g\}.$$

And **(d)** is expressed as:

$$f/g = \{(s, t/t') : (s, t) \in f, \ (s, t') \in g, \ t' \neq 0\}.$$

17. Take:

$$f(n) = \begin{cases} 0 & \text{if } n = 1 \\ n/2 & \text{if } n \text{ is even} \\ -(n-1)/2 & \text{if } n \text{ is odd and } n \geq 3 \end{cases}$$

19. Suppose that we had such a function g. Construct some subset A of $\mathbb{N}$ where g is injective, i.e., choose one representative from the preimage of each point x in $\mathbb{R}$. Then the restriction of g to A will give a bijection from $A \subset \mathbb{N}$ into $\mathbb{R}$. That would imply that the cardinality of $\mathbb{N}$ is greater than or equal to the cardinality of $\mathbb{R}$. We know that that is not so by Example 4.5.21 in the text.

20. (c) Countable, by Proposition 4.5.15.

   (d) Uncountable (same cardinality as $\mathbb{R}$).

   (f) Uncountable.

   (g) Countable, since it is contained in $\mathbb{Q}$.

   (i) Countable, since they are contained in $\mathbb{Q}$.

   (k) Uncountable, since any real number can be such a solution.

   (m) Uncountable, since we are taking away just a small piece of $\mathcal{P}(\mathbb{Z})$,

23. We can rewrite the property of f as follows: for each $x \in \mathbb{R}$ there is an interval (p_x, q_x) with $p_x, q_x \in \mathbb{Q}$ and $x \in (p_x, q_x)$ such that $f(t) \geq f(x)$ for all $t \in (p_x, q_x)$. Now let y be a point in the image of f. Choose x_y such that $f(x_y) = y$. Then there must be an interval (p_{x_y}, q_{x_y}) with rational endpoints, containing x and satisfying $f(t) \geq f(x)$ for all $t \in (p_x, q_x)$. Hence, for each point y in the image of f we can assign an element $(p_{x_y}, q_{x_y}) \in \mathbb{Q} \times \mathbb{Q}$. This assignment is one-to-one: if y and y' were assigned the same interval, we would have $x_{y'} \in (p_{x_y}, q_{x_y})$ and $x_y \in (p_{x'_y}, q_{x'_y})$. Thus,

$$y' = f(x_{y'}) \geq f(x_y) = y \text{ and } y = f(x_y) \geq f(x_{y'}) = y'.$$

Therefore we must have $y = y'$. Thus, the assignment $y \rightarrow (p_{x_y}, q_{x_y}) \in \mathbb{Q} \times \mathbb{Q}$ is one-to-one and has a countable target, which implies that the image of f is countable.

25. (a) Domain : Nonnegative reals.
Range : Real numbers greater or equal than -3.
(c) Domain : Males with offsprings.
Range : Human beings (everybody has, or had, a father).
(e) Domain : $(-1, 1) \cap \mathbb{Q}$.
Range : $(-1, 1) \cap \mathbb{Q}$.
(g) Domain : $\mathbb{R}$.
Range : $\mathbb{R}$.

27. We can define an ordered triple as:

$$(a, b, c) \leftrightarrow \{\{a\}, \{a, b\}, \{\{a, b\}, c\}\}.$$

30. Domain of f : $\{1, 2, 3\}$.
Image of f : $\{a, c\}$.
Domain of g : $\{a, b, c\}$.
Image of g : $\{t, s, u\}$.
$g \circ f = \{(1, u), (2, u), (3, t)\}$.
$g^{-1} = \{(t, a), (s, b), (u, c)\}$.

31. If a function $f : X \to Y$ is defined as a subset of $X \times Y$, we can say that the intersection of two functions is the intersection of the subsets of $X \times Y$ that define them. To see that this is still a function, note that if $(x, y) \in f \cap g$ and $(x, y') \in f \cap g$, then $(x, y) \in f$ and $(x, y') \in f$, so we must have $y = y'$. Therefore, $f \cap g$ is a function (with smaller domain unless $f = g$).

Concerning the union, if we define $f \cup g$ as the union of the subsets of $X \times Y$ that define them, then whenever $y = f(x) \neq g(x) = y'$, we could have $(x, y) \in f \cup g$ and $(x, y') \in f \cup g$, and $y \neq y'$, so $f \cup g$ is not a function in general.

33. Take $y \in \mathbb{R}$. Then construct $g : \mathbb{R} \to \mathbb{Z}$ by taking $\mathbb{R} \setminus \{y\}$ onto $\mathbb{Z} \setminus \{0\}$ (do this by sending each real number to its integer part), and y to 0. Then by hypothesis $g \circ f$ is onto, so 0 is in the image of $g \circ f$. Since the only element that can be taken to 0 by g is y, it must be that y is in the image of f. Since y was arbitrary, f must be onto.

35. Do the same process as in the previous exercise. The set K that we obtain is uncountable, since it has the same cardinality as $\mathbb{R}$. But $K \subset X$, so X must be uncountable as well.

37. It is not an order relation: it does not satisfy property (a) of partial ordering and it cannot be a total ordering since not every two elements are comparable (for example, 1 and 2 are not comparable).

39. This is an equivalence relation. Note that 0 is a root of $p - q$ if and only if $p(0) = q(0)$, which is true if and only if x divides $p - q$. It is now routine to check that reflexivity, symmetry, and transitivity hold.

41. The only thing to be checked here is that the function constructed in the text is an isomorphism. But this is clear: it is an isomorphism in U (because it is just the identity) and it is an isomorphism between T and T' (because g was assumed to be an isomorphism). Since T and U are disjoint and their union is S, the function is well defined and it is an isomorphism.

43. **Reflexive :** $A\mathcal{R}A$ since the identity is a set theoretic isomorphism of A with itself.

    Symmetric : If f is a set theoretic isomorphism from A to B, then f^{-1} is a set theoretic isomorphism from B to A.

    Transitive : If f and g are set theoretic isomorphisms from A to B and from B to C, respectively, then $g \circ f$ is a set theoretic isomorphism from A to C.

45. It is routine to check that it is an equivalence relation. Each equivalence class in each interval with integer endpoints consists of numbers that differ from each other by less than 10^{-5}. In fact, the set of equivalence classes can be identified with the set of intervals of the form $[0, 10^{-5}), [10^{-5}, 2 \cdot 10^{-5}), [2 \cdot 10^{-5}, 3 \cdot 10^{-5})$, etc., contained in $[0, 1)$.

50. For each of the m elements of T there are k possible assignments of elements of S. This simply says that there are

$$\underbrace{k \cdot k \cdots \cdot k}_{m \text{ times}}$$

    possible functions from T to S.

51. Write $T = \{1, 2, 3\}$. The function

$$\begin{aligned} S^T &\longrightarrow S \times S \times S \\ f &\longmapsto (f(1), f(2), f(3)) \end{aligned}$$

    is plainly one-to-one and onto.

53. There are no functions from $\emptyset$ to S. So S^T is empty. Likewise, T^S is empty.

55. Let $S = \{0, 1\}^{\mathbb{N}}$. If $f \in S$, then associate to f the sequence $\{f(1), f(2), \dots\}$. The function f is clearly one-to-one and onto.

Chapter 5: Number Systems

1. $\mathbb{Q}$ is closed under all of these operations except $\div$ (only because we cannot divide by 0). $\mathbb{R} \setminus \mathbb{Q}$ is not closed under any of these operations. (For example, $\sqrt{2} - \sqrt{2} = 0 \in \mathbb{Q}$, $\sqrt{2} \cdot \sqrt{2} = 2 \in \mathbb{Q}$.)

2. Do the same as in the previous exercise, using addition instead of multiplication (cf. Exercise 4.3). One can choose the representative $(n', 1)$ from each class and realize that addition and multiplication of classes lead to the usual addition and multiplication of n''s as rational numbers. So the set of equivalence classes is actually the set of rational numbers with which we began.

5. Notice that $x + y = (\pi + e) + (\pi - e) = 2\pi$, which is irrational. If both $(\pi + e)$ and $(\pi - e)$ were rational, their sum would be too.

8. We have that $\{a_j\}_{j=1}^{\infty}$ lies between $-M$ and M. We can assume that M is a positive integer such that $|a_j| < M - 1$, i.e., $M - 1$ is also an upper bound (note that there are many upper bounds to choose from, so we can certainly do this). Let X be the set defined as:

$$X = \{t \in [-M, M] \text{ with only finitely many } a_j\text{'s in } (t, M] \}.$$

This set is bounded, since it is contained in $[-M, M]$. It is also non-empty, since $M - 1 \in X$. Therefore this set must have a greatest lower bound x. Given $\epsilon > 0$, note first that there must be only finitely many a_j's in the interval $[x + \epsilon, M]$, for otherwise x would not be a lower bound of X. But we also know that there are infinitely many a_j's in $(x - \epsilon, M]$, for otherwise $x - \epsilon$ would be a lower bound which is less than x. So it must be that there are infinitely many a_j's in the interval $(x - \epsilon, x + \epsilon)$. In particular, for any $N > 0$, there is $j > N$ such that $a_j \in (x + \epsilon, x - \epsilon)$, i.e., $|a_j - x| < \epsilon$.

9. If $(x, y) \sim (x', y')$ and $(w, z) \sim (w', z')$, then

$$
\begin{aligned}
(x + w) + (y' + z') &= (x + y') + (w + z') \\
&= (x' + y) + (w' + z) \\
&= (x' + w') + (y + z),
\end{aligned}
$$

which shows that $(x + w, y + z) \sim (x' + w', y' + z')$.

13. Reflexivity and symmetry are clear. To prove transitivity, suppose that $\{a_j\}$ is related to $\{a'_j\}$ and that $\{a'_j\}$ is related to $\{a''_j\}$. Take N so that $|a_j - a'_k| < \epsilon/2$ if $k, j > N$ and M so that $|a'_j - a''_k| < \epsilon/2$ if $k, j > M$. Then

$$|a_j - a''_k| \leq |a_j - a'_k| + |a'_j - a''_k| < \epsilon \text{ if } k, j > \max\{M, N\}.$$

The real numbers can be thought of as the "set of limits" of rational numbers. The equivalence relation above is the same as the relation "having the same limit." In this way, for each class of sequences, one can assign a real number which is, in the end, the limit of that class. One can also assign a unique Dedekind cut to each Cauchy sequence, using the recipe:

$$\bigcup \{x \in \mathbb{Q} \text{ such that only finitely many } a_j \text{ are less than } x\}.$$

14. If $(x, y) \sim (x', y')$ and $(w, z) \sim (w', z')$, then

$$
\begin{aligned}
(xz + wy)y'z' &= xy'zz' + wz'yy' \\
&= x'yzz' + w'zyy' \\
&= (x'z' + w'y')yz,
\end{aligned}
$$

which is what we need to prove for well-definedness of addition. Subtraction works similarly.

15. Let $x = \sqrt{2} + \sqrt{3}$. Suppose that x is rational. Then x^2 is rational, so $2 + 2\sqrt{6} + 3$ is rational. Hence $\sqrt{6}$ is rational. But that is not true. Hence $x = \sqrt{2} + \sqrt{3}$ is irrational.

16. Addition and subtraction of cuts are clearly well defined (there is nothing to check since there are no choices involved in the definition). Each real number corresponds to a cut in a bijective way. Thus, addition and subtraction of reals must also be well defined.

17. For exactly the same reason as in the previous two exercises: there are no choices involved in the definition.

19. Let $[(a, b)], [(a', b')], [(a'', b'')]$ be integers, written as equivalence classes of pairs of natural numbers. We have to show that

$$
([(a, b)] + [(a', b')]) + [(a'', b'')] = [(a, b)] + ([(a', b')] + [(a'', b'')]).
$$

This can be done as follows:

$$
\begin{aligned}
&([(a, b)] + [(a', b')]) + [(a'', b'')] \\
&= [(a + a', b + b')] + [(a'', b'')] \\
&= [((a + a') + a'', (b + b') + b'')] \\
&= [(a + (a' + a''), b + (b' + b''))] \\
&= [(a, b)] + [(a' + a'', b' + b'')] \\
&= [(a, b)] + ([(a', b')] + [(a'', b'')]),
\end{aligned}
$$

where we used associativity of addition of natural numbers in the third equality.

21. An Argand diagram looks like this:

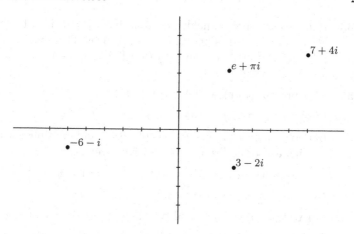

23. The roots are $1, -1/2 + i\sqrt{3}/2, -1/2 - i\sqrt{3}/2$. They can be found using the formulas in Exercise 6.26 and Exercise 6.27 below. They form the vertices of an equilateral triangle inscribed in the circle of radius one around the origin.

27. One only has to write things out:

$$
\begin{aligned}
re^{i(\theta+2\pi)} &= r(\cos(\theta + 2\pi) + i\sin(\theta + 2\pi)) \\
&= r(\cos(\theta) + i\sin(\theta)) \\
&= re^{i\theta}.
\end{aligned}
$$

28. The kth roots of a number $z = re^{i\theta}$ are:

$$
r^{1/k}e^{i\theta/k}, r^{1/k}e^{i(\theta+2\pi)/k}, \ldots, r^{1/k}e^{i(\theta+2(k-1)\pi)/k}.
$$

Using this formula, we find that the six sixth roots of 2 are:

$$
\sqrt[6]{2},\ \sqrt[6]{2}\,e^{\frac{\pi i}{3}},\ \sqrt[6]{2}\,e^{\frac{2\pi i}{3}},\ \sqrt[6]{2}\,e^{\pi i},\ \sqrt[6]{2}\,e^{\frac{4\pi i}{3}},\ \sqrt[6]{2}\,e^{\frac{5\pi i}{3}}.
$$

On an Argand diagram, the six roots of 2 form the vertices of an hexagon inscribed in a circle of radius $\sqrt[6]{2}$.

29. $1 + i = \sqrt{2}e^{\pi/4}$, so the fourth roots are:

$$
\sqrt[8]{2}\,e^{i\pi/16},\ \sqrt[8]{2}\,e^{i(\pi/16+\pi/2)},\ \sqrt[8]{2}\,e^{i(\pi/16+\pi)},\ \sqrt[8]{2}\,e^{i(\pi/16+3\pi/2)}
$$

In the Argand plane, the four roots form the vertices of a square of side 2.

32. We have addition and multiplication defined in both $\mathbb{C}$ and $\mathbb{H}$. What the statement in the exercise means is that, by the given correspondence, the sum and the product of two complex numbers goes to the sum and the product, respectively, of the corresponding quaternions (cf. Exercise 6.41).

33. If it were an ordered field, we would have $1 > 0$ since $1 = 1^2 > 0$ by Theorem 6.13 (5). This implies $-1 < 0$ by Theorem 6.13 (3). But $-1 = i^2 > 0$ by Theorem 6.13 (5). This gives $-1 < 0$ and $-1 > 0$, a contradiction.

35. **(2)** Apply part **(1)** to $x + z = 0 = x + (-x) = 0$.

(3) Apply part **(2)** to the expression $z + (-z) = 0$.

(4) Do as in part **(1)** using multiplication and multiplicative inverses instead of addition and additive inverses.

(5) Apply part **(4)** to the equation $y \cdot z = 1 = y \cdot y^{-1}$.

(6) Apply part **(5)** to the expression $y^{-1} \cdot y = 1$.

(9) Add $(x \cdot y)$ to the equality. Then apply D1 and A5.

(10) Substitute y^{-1} for y in part **(9)** and then apply part **(3)**.

37. Take $\epsilon = \frac{y^2 - x}{4(x+1)}$. We have to show that

$$\left(y - \frac{y^2 - x}{4(x+1)}\right)^2 \geq x.$$

Expanding the square, we find that

$$\left(y - \frac{y^2 - x}{4(x+1)}\right)^2 = y^2 - 2y\frac{y^2 - x}{4(x+1)} + \left(\frac{y^2 - x}{4(x+1)}\right)^2.$$

Now notice that $y < 2(x+1)$, so

$$-\frac{y}{2(x+1)} > -1.$$

Using this and the fact that the last term in the long equation above is positive, we finally get

$$\begin{aligned}\left(y - \frac{y^2 - x}{4(x+1)}\right)^2 &= y^2 - 2y\frac{y^2 - x}{4(x+1)} + \left(\frac{y^2 - x}{4(x+1)}\right)^2 \\ &> y^2 - (y^2 - x) + \left(\frac{y^2 - x}{4(x+1)}\right)^2 \\ &> y^2 - (y^2 - x) \\ &= x\end{aligned}$$

39. Write the required identity in terms of pairs of real numbers. Then use commutativity of addition and multiplication of reals to obtain the result.

40. Writing complex numbers as ordered pairs of complex numbers, we can compute:

$$
\begin{aligned}
(a,b) &\cdot [(c,d) + (c',d')] \\
&= (a,b) \cdot (c+c', d+d') \\
&= (a(c+c') - b(d+d'), a(d+d') + b(c+c')) \\
&= ((ac-bd) + (ac'-bd'), (ad+bc) + (ad'+bc')) \\
&= (ac-bd, ad+bc) + (ac'-bd', ad'+bc') \\
&= (a,b) \cdot (c,d) + (a,b) \cdot (c',d').
\end{aligned}
$$

41. One way is via the mapping

$$ x + iy \longmapsto x\mathbf{1} + y\mathbf{i} + 0\mathbf{j} + 0\mathbf{k}. $$

Another is via the mapping

$$ x + iy \longmapsto x\mathbf{1} + 0\mathbf{i} + y\mathbf{j} + 0\mathbf{k}. $$

43. $|z \cdot \bar{z}| = |z|^2$. This is exactly what happens with the complex numbers.

45. We know that we can write

$$ p(z) = d(z) \cdot (z - \alpha) + r, $$

(just do long division), where r is a polynomial of degree less than 1—hence a constant. Now plug in $z = \alpha$. The left-hand side gives 0 because α is a root. The right-hand side gives r. Thus, $r = 0$, and we have

$$ p(z) = d(z) \cdot (z - \alpha). $$

49. (2)Write $z = a + bi$, $w = c + di$. Then

$$
\begin{aligned}
\overline{z \cdot w} &= \overline{(a+bi) \cdot (c+di)} \\
&= \overline{(ac-bd) + i(ad+bc)} \\
&= (ac-bd) - i(ad+bc) \\
&= (a-bi) \cdot (c-di) \\
&= \overline{(a+bi)} \cdot \overline{(c+di)}.
\end{aligned}
$$

(3) $z + \bar{z} = (a+ib) + (a-ib) = 2a = 2 \cdot \operatorname{Re} z.$

(4) $z - \bar{z} = (a+ib) - (a-ib) = 2ib = 2i \cdot \operatorname{Im} z.$

50. (1) $|z| = \sqrt{a^2 + b^2} = \sqrt{a^2 + (-b)^2} = |\bar{z}|.$

(2)Using $|z|^2 = z\bar{z}$, we find:

$$
\begin{aligned}
|z \cdot w| &= \sqrt{\overline{z \cdot w} \cdot z \cdot w} \\
&= \sqrt{z\bar{z}w\bar{w}} \\
&= \sqrt{z\bar{z}}\sqrt{w\bar{w}} \\
&= |z| \cdot |w|.
\end{aligned}
$$

(3)$|\mathrm{Re}\, z| = |a| = \sqrt{a^2} \le \sqrt{a^2 + b^2} = |z|$.

$|\mathrm{Im}\, z| = |b| = \sqrt{b^2} \le \sqrt{a^2 + b^2} = |z|$.

53. Let $z = i\pi/2$ and let $w = -i\pi/2$.

Chapter 6: More on the Real Number System

1. Let $\lambda = \eta/(1 + 2\eta)$. Now construct a Cantor set by removing a middle open interval of length λ, then two side intervals of length λ^2, then four intervals of length λ^3, etc. The sum of the lengths of the removed intervals will be η, as desired.

3. There are $1 + 2 + 4 + \cdots$, hence countably many removed intervals. Each has two endpoints. So there are only countably many endpoints.

5. Let $\mathcal{O}$ be the collection of interior points of S. Select $p \in \mathcal{O}$. By definition, there is an $\epsilon > 0$ such that $(p - \epsilon, p + \epsilon) \subset S$. If $t \in (p - \epsilon, p + \epsilon)$, then set $\eta = \min\{t - (p - \epsilon), t - (p + \epsilon)\}$. By the triangle inequality, the interval $(t - \eta, t + \eta) \subset (p - \epsilon, p + \epsilon) \subset \mathcal{O}$. This shows that $\mathcal{O}$ is open.

7. Let x be any point of S. Either **(i)** there exists an $\epsilon > 0$ such that $(x - \epsilon, x + \epsilon) \subset S$ or else **(ii)** for each $\epsilon > 0$, both $(x - \epsilon, x + \epsilon) \cap S \neq \emptyset$ and $(x - \epsilon, x + \epsilon) \cap {}^c S \neq \emptyset$. In the first of these cases, x is in the interior of S; in the second case x is in the boundary of S.

9. The set $\mathbb{Q}$ is not discrete. For if $q \in \mathbb{Q}$ and $\epsilon > 0$, then choose a positive integer N such that $1/N < \epsilon$. Then $r = q + 1/N$ is rational and $|q - r| = 1/N < \epsilon$. The Cantor set is also not discrete. In fact it is *perfect* (Exercise 7.4), so each element is the limit of a sequence of other elements.

   The set $\mathbb{Z}$ of integers *is* discrete. If $k \in \mathbb{Z}$, then let $\epsilon = 1/2$. Then $(k - 1/2, k + 1/2) \cap \mathbb{Z} = \{k\}$.

   The set $T = \{1, 1/2, 1/3, \dots\}$ is discrete. For if $1/j \in T$, then let $\epsilon = |(1/j) - (1/[j + 1])|$. Then $(1/j - \epsilon, 1/j + \epsilon) \cap T = \{1/j\}$.

11. Let $S = \{s_j\}$. If S is bounded above, then let α be a least upper bound for S. Then it is easy to show that $s_j \to \alpha$. If S is *not* bounded above, then we may choose $s_{j_1} > 1, s_{j_2} > 2$, and so forth. So S is unbounded.

13. Let $\{\ell_j\}$ be a sequence of limit points with $\ell_j \to \alpha$. We need to show that $\alpha \in L$. There is a point $s_1 \in S$ such that $|s_1 - \ell_1| < 1/1$. Then there is a point $s_2 \in S$ such that $|s_2 - \ell_2| < 1/2$. Continue in this manner. It follows from the triangle inequality that the sequence $\{s_j\}$ converges to α. So α is a limit point as required.

15. Let $x \in \mathcal{C} \setminus \mathcal{O}$. Then for each $\epsilon > 0$ it is the case that $(x - \epsilon, x + \epsilon) \cap {}^c S \neq \emptyset$. If also $(x - \epsilon, x + \epsilon) \cap S \neq \emptyset$ then, by definition, x is in the boundary of S. It cannot be the case that there is some $\epsilon > 0$ such that $(x - \epsilon, x + \epsilon) \cap S = \emptyset$ because then x would not even be a limit point of S. Thus we see that x must be in the boundary of S. In conclusion, $\mathcal{C} \setminus \mathcal{O} \subset$ (boundary of S). The converse is straightforward.

19. If $x_1 = 1$, $x_2 = 1/2$, $x_3 = 1/3$, etc., then $X \setminus \{x_j\} = \{0\}$. By contrast, if $\{q_j\}$ is an enumeration of the rationals and X is the closure of that set (which is, of course, the reals), then $X \setminus \{q_j\}$ is the set of all irrational numbers—an uncountable set.

20. If $x \in \mathbb{Q}$ and $\epsilon > 0$, then $(x - \epsilon, x + \epsilon) \not\subset \mathbb{Q}$. So x is *not* in the interior of $\mathbb{Q}$. Thus $\mathbb{Q}$ has no interior. A similar argument shows that the Cantor set has no interior. On the other hand, if $x \in \mathbb{R}$ and $\epsilon = 1/2$, then $(x - 1/2, x + 1/2) \subset \mathbb{R}$. So x is in the interior of $\mathbb{R}$. We conclude that the interior of $\mathbb{R}$ is $\mathbb{R}$ itself.

21. If $x \in \cup_j \mathcal{O}_j$, then $x \in \mathcal{O}_{j_0}$ for some j_0. Hence there is an $\epsilon > 0$ such that $(x - \epsilon, x + \epsilon) \subset \mathcal{O}_{j_0}$. It follows then that $(x - \epsilon, x + \epsilon) \subset \cup_j \mathcal{O}_j$. So $\cup_j \mathcal{O}_j$ is open.

By contrast, if $\mathcal{O}_j = (-1/j, +1/j)$, then each $\mathcal{O}_j$ is open but $\cap_j \mathcal{O}_j = \{0\}$, which is not open (in fact it is closed).

We may derive the facts about closed sets simply by complementation. The intersection of a collection of closed sets is definitely closed. But the union of closed sets need not be closed.

23. Let $U \subset \mathbb{R}$ be a set. Suppose that whenever $\{x_j\} \subset {}^c U$ is a sequence and $x_j \to \alpha \in \mathbb{R}$, then $\alpha \in {}^c U$. Then U is open. The reason, of course, is that this gives a sequential characterization of the complement of U being closed.

25. Write $\mathbb{R} = \cup_j [j, j+1]$. Then one of these intervals, call it $[j_0, j_0 + 1]$, must contain uncountably many elements of S. Now write $[j_0, j_0 + 1] = [j_0, j_0 + 1/2] \cup [j_0 + 1/2, j_0 + 1]$. One of the two subintervals must contain uncountably many elements of S. Bisect that interval and choose the half that contains uncountably many elements.

Continue. Then choose one element from each of the sequence of chosen intervals. These form a sequence that clearly converges to the point that is the intersection of the chosen intervals. So there exists at least one limit point of S.

If instead S is countable, then the assertion is false. If $S = \mathbb{Z}$, then S has no limit points.

27. Define $\mathcal{O}_j = \{x \in \mathbb{R} : \text{dist}(x, E) < 1/j\}$. Then each $\mathcal{O}_j$ is clearly open, and $\mathcal{O}_1 \supset \mathcal{O}_2 \supset \cdots$. Finally, $\cap_j \mathcal{O}_j = E$.

28. Let $E = \{0\}$. Then E is closed, but E cannot be written as the increasing union of open sets. In any event, the increasing union of open sets is open (Exercised 7.21). Thus we cannot switch "open" and "closed" in Exercise 7.26.

Likewise, the decreasing intersection of closed sets is closed (Exercise 7.21). So we cannot switch "open" and "closed" in Exercise 7.27.

29. Let $D(P, r)$ denote the disc in the plane with center P and radius r. Say that a set $U \subset \mathbb{R}^2$ is open if, for each $P \in U$, there is an $r > 0$ such that $D(P, r) \subset U$.

If V, W are open subsets of $\mathbb{R}$, then let $X = (x, y)$ be a point of $V \times W$. Since V is open, there is an $r_1 > 0$ such that $(x - r_1, x + r_1) \subset V$. And there is an r_2 such that $(y - r_2, y + r_2) \subset W$. Therefore

$$(x - r_1, x + r_1) \times (y - r_2, y + r_2) \subset V \times W.$$

Let $r_0 = \min(r_1, r_2)$. Then

$$D(X, r_0) \subset (x - r_1, x + r_1) \times (y - r_2, y + r_2) \subset V \times W.$$

So $V \times W$ is open.

30. This follows from Exercise 29 by complementation.

Chapter 7: Elementary Number Theory

1. We use the Division Algorithm as in the text:

$$
\begin{aligned}
1235 &= 435 \cdot 2 + 365 \\
435 &= 365 \cdot 1 + 70 \\
365 &= 70 \cdot 5 + 15 \\
70 &= 15 \cdot 4 + 10 \\
15 &= 10 \cdot 1 + 5 \\
10 &= 5 \cdot 2 + 0.
\end{aligned}
$$

We conclude that the greatest commond divisor is 5.

3. As in the text. Suppose that $p_1, p_2, \ldots, p_k$ are all the primes of the form $6x - 1$. Consider the number

$$N = 6p_1p_2 \cdots \cdot p_k - 1.$$

Certainly $p_j \nmid N$ for each $j = 1, \ldots, k$. Not every prime divisor of N is of the form $6x + 1$; if they all were, then N itself would be of this form but it is not. For similar reasons, not every prime divisor of N is of the form $6x + 3$. Also, not every prime divisor of N is of the form $6x + 5$. If the prime divisors of N are a mixture of numbers of the form $6x + 1$, $6x + 3$, and $6x + 5$, then N must be of the form $6x + 3$, which it is not. We conclude that N has a prime divisor of the form $6x - 1$, and that number is perforce different from each of the $p_1, p_2, \ldots, p_k$. So we have produced a new prime of the form $6x - 1$, and that is a contradiction.

5. (a) Every prime factor of a divides n and every prime factor of b divides n. And the prime factors of a are distinct from the prime factors of b. So the product of *all* the prime factors divides n. Hence $ab|n$.

(b) Each prime factor of a occurs in the integer bc. But b is relatively prime to a, so none of these prime factors occurs in b. The all occur in c. Thus $a|c$.

6. (a) Every prime factor of a occurs in b and every prime factor of b occurs in c. Hence every prime factor of a occurs in c. Thus $a|c$.

(c) Every prime factor of a occurs in b. Thus every prime factor of ma occurs in mb. So $ma|mb$.

7. (a)

$$300 = 17 \cdot 17 + 11$$

(d)

$$389 = 97 \cdot 4 + 1$$

8. If $a^n|b^n$ then every prime factor of a divides b^n so every prime factor of a divides b. Hence $a|b$.

9. Every positive integer is of one of the forms $4m$, $4m + 1$, $4m + 2$, or $4m + 3$. Now

$$(4m)^2 \mod 4 = 0$$
$$(4m + 1)2 \mod 4 = 1$$
$$(4m + 2)^2 \mod 4 = 0$$
$$(4m + 3)^2 \mod 4 = 1.$$

So we see that no perfect square has residue 3 modulo 4. That proves the result.

10. This is immediate from the hint and Exercise 9.9.

11. Denote $[\mathbf{Z}/n\mathbf{Z}]^*$ by G. Clearly $1 \in G$ and it is the multiplicative identity. Also, if j, k are relatively prime to n then $j \cdot k$ is also relatively prime to n. So G is closed under multiplication. If $j_1, j_2, \ldots, j_k$ is an enumeration of the elements of G and if m is any element of G, then $m \cdot j_1, m \cdot j_2, \ldots, m \cdot j_k$ are distinct. So one of these must be 1. Hence m has a multiplicative inverse. The associativity of multiplication is obvious. We conclude that G is a group.

12. (a)

$$15 = 5 \cdot 3 + 0.$$

We conclude that 5 is the greatest common divisor.

(c)

$$
\begin{aligned}
897 &= 51 \cdot 17 + 30 \\
51 &= 30 \cdot 1 + 21 \\
30 &= 21 \cdot 1 + 9 \\
21 &= 9 \cdot 2 + 3 \\
9 &= 3 \cdot 3 + 0.
\end{aligned}
$$

We conclude that 3 is the greatest commond divisor.

14. We see that

$$(a+b)^p = a^p + \binom{p}{1} a^{p-1} b + \binom{p}{2} a^{p-2} b + \cdots + b^p.$$

Now each of these binomial coefficients is divisible by p. Hence

$$(a+b)^p \mod p = (a^p + b^p) \mod p.$$

16. Divide the equation through by 17. The result is

$$3x + 7y = 1.$$

It is easy to see that $x = -2$, $y = 1$ solves the equation.

17. Write

$$n = a + 10b + 100c + \cdots,$$

where the digits of n are a, b, c, etc. Since $10 = 1 \mod 3$, we find that

$$n = a + 10b + 100c + \cdots \mod 3 = a + b + c + \ldots \mod 3.$$

That is the result.

21. Clearly $\mathbf{Z}/p\mathbf{Z}$ is closed under addition. If $m, n \in \mathbf{Z}/p\mathbf{Z}$, then $m \cdot n$ mod p makes sense and lies in $\mathbf{Z}/p\mathbf{Z}$. The multiplicative identity is 1. The elements of $\mathbf{Z}/p\mathbf{Z}$ are $0, 1, 2, \ldots, (p-1)$. If m is any element of the space other than 0, then $m \cdot 0, m \cdot 1, \ldots, m \cdot (p-1)$ are distinct. So one of them must be 1. Thus m has a multiplicative inverse. The other properties of a field are obvious.

23. Refer to the solutions of Exercises 9.11 and 9.21.

25. Calculating powers of 3, we see that the last digits are $3, 9, 7, 1, 3, 9, 7, 1, \ldots$. The four-cycle $3, 9, 7, 1$ repeats indefinitely. Since $45 = 1 \mod 4$, we conclude that the last digit of 3^{45} is 3.

Chapter 8: Zero-Knowledge Proofs and Cryptography

1. First, we translate the message (ignoring spaces) into digits:

   19 07 04 01 04 00 19 11 04 18 05 14 17 04 21 04 17 .

   We add three Xs to the end of the message so that the number of numbers we have is a multiple of 5. So in fact our list of digits is

   19 07 04 01 04 00 19 11 04 18 05 14 17 04 21
   04 17 23 23 23 .

   The next step is to break this list up into 5-number "words:"

   19 7 04 01 04   00 19 11 04 18   05 14 17 04 21
   04 17 23 23 23 .

   Now we apply the RSA technology to each of the "words"

   1907040104   0019110418   0514170421   0417232323 .

   Thus we obtain

   $$w_1 = (1907040104)^{1225} \mod 34161,$$

   $$w_2 = (0019110418)^{1225} \mod 34161,$$

   $$w_3 = (0514170421)^{1225} \mod 34161,$$

   $$w_4 = (0417232323)^{1225} \mod 34161,$$

   These four numbers—which you may calculate if you like—constitute the RSA encryption of "The Beatles forever."

4. Certainly, if m_1 is an English message transliterated to a string of integers with the usual association of numbers with letters, and if m_2 is another such message transliterated, then

$$(m_1)^e \bmod n \cdot (m_2)^e \bmod n = (m_1 \cdot m_2) \bmod n.$$

So the messages may be encrypted separately and then multiplied, or they can be multiplied and then encrypted. This device can sometimes be used to crack a message more quickly.

5. With the Chinese Remainder Theorem, and with $n = p \cdot q$ as in RSA encryption, one can calculate

$$w^e \bmod n$$

sometimes more effectively by calculating

$$w^e \bmod p$$

and

$$w^e \bmod q$$

and then combining.

7. ALL THE WORLD IS A STAGE AND ALL THE MEN AND WOMEN MERELY PLAYERS.

9. The Playfair Cipher encryption is

RFDLBAFXZVAVMA

11. We calculate

$$
\begin{aligned}
248 &= 1 \cdot 164 + 84 \\
164 &= 1 \cdot 84 + 80 \\
84 &= 1 \cdot 80 + 4 \\
80 &= 20 \cdot 4 + 0.
\end{aligned}
$$

The calculation ends when the remainder is 0. The last divisor is 4, and that is the greatest common divisor.

14. We see that
$$4^{13-1} = 4^{12} = 16777216.$$

And
$$16777216 = 13 \cdot 1290555 + 1.$$

Hence
$$16777216 \bmod 13 = 1.$$

So 13 passes the test for primality

We also see that

$$5^{13-1} = 5^{12} = 244140625.$$

And

$$244140625 = 13 \cdot 18780048 + 1.$$

Hence

$$244140625 \bmod 13 = 1.$$

Once again, 13 passes the test.

This is no surprise because, after all, 13 is a prime integer.

15. Let $p = 14$ and $a = 3$. We calculate that

$$a^{p-1} = 1594323.$$

Furthermore,

$$a^{p-1} \bmod p = 3.$$

So this p fails, and we conclude that p is composite.

Let $p = 14$ and $a = 4$. We calculate that

$$a^{p-1} = 67108864.$$

Furthermore,

$$a^{p-1} \bmod p = 4.$$

So this p fails, and we conclude that p is composite.

Let $p = 14$ and $a = 5$. We calculate that

$$a^{p-1} = 1220703125.$$

Furthermore,

$$a^{p-1} \bmod p = 5.$$

So this p fails, and we conclude that p is composite.

17. Of course $p^2 - q^2 = (p - q) \cdot (p + q)$, so that $p^2 - q^2$ is composite. But if $p = 2$ and $q = 3$ then $p^2 + q^2 = 13$, which is prime.

19. Let p be the product of the p_j. Set

$$y_j = (p/p_j)^{\varphi(p_j)} \quad \text{for } j = 1, \ldots, k.$$

Here φ is, as usual, the Euler φ function (i.e., it counts the number of integers less than the argument that are relatively prime to the argument). Then

$$y_j \equiv 1 \bmod p_j \quad \text{for } j = 1, \ldots, k.$$

This last follows from Fermat's Little Theorem.

Finally, set

$$x = a_1 y_1 + a_2 y_2 + \cdots + a_k y_k.$$

Chapter 9: Examples of Axiomatic Theories

1. Define the group operation of $G \times H$ as:

$$(g, h) \cdot (g', h') = (g \cdot g', h \cdot h').$$

It is not difficult to check that this operation gives $G \times H$ a group structure. Basically, one has to write out the required properties and then apply the corresponding properties given by the group structures of G and H. The identity element will be (e_G, e_H), where e_G, and e_H are the identity elements of G and H, respectively. The inverse of an element (g, h) will be (g^{-1}, h^{-1}), where g^{-1} and h^{-1} are the inverses of g and h in G and H, respectively.

If G and H are abelian, $G \times H$ will also be abelian, for

$$(g, h) \cdot (g', h') = (g \cdot g', h \cdot h') = (g' \cdot g, h' \cdot h) = (g', h') \cdot (g, h).$$

3. The only one is $\mathbb{Z}_2$.

5. There are two different groups: $\mathbb{Z}_2 \times \mathbb{Z}_2$ and $\mathbb{Z}_4$.

10. For the first definition of T: Observe that $-1 \in T, 3 \in T$, but $-1 + 3 = 2 \notin T$. Hence T is not closed under addition. Also $-1 \in T, 3 \in T$, but $(-1) \cdot 3 = -3 \notin T$. Hence T is not closed under multiplication.

For the second definition of T: The set T consists of all multiples of 2. Multiples of 2 are closed both under addition and multiplication.

11. The assertion means that the fifth axiom is independent of the others. In other words, if the first four axioms are compatible with each other, the system will still be compatible if we add either Axiom 5 or its negation to the previous four, because the fifth axiom is independent of the others.

13. Any cyclic group G is abelian, so any subgroup H of G is automatically normal. Therefore G/H is a group. Now we have to prove that it is cyclic. If $g \in G$ generates G (i.e., every element of G is of the form g^k for some k), then any element of G/H can be written as $g^k H$. But $g^k H = (gH)^k$, so $(gH)^k$ is a generator of G/H.

15. In hyperbolic geometry there are no similar triangles, i.e., the angles of a triangle determine its sides uniquely. One can therefore develop formulas relating the sides with the angles. For example, for a triangle $\triangle ABC$ one has:

$$\sin(\angle CAB) = \frac{\sinh(CB)}{\sinh(AB)} \quad \text{and} \quad \cos(\angle CAB) = \frac{\tanh(CA)}{\tanh(AB)},$$

where the quantities CB, AB, CA are the lengths of the sides of the triangle measured in the Poincaré sense.

A detailed account of how to derive these formulas is given in [GRE], p. 329 ff.

17. The points are points in the open upper half plane (not including the x-axis). The lines are either vertical straight lines or semicircles that intersect the x-axis perpendicularly. The distance between two points gets larger as we approach the x-axis. "There is an infinite distance between any point and the x-axis" (written between quotation marks because the points in the x-axis are not really points in our geometry; the idea to understand is that the plane extends indefinitely as we approach the x-axis).

20. Consider the set of those matrices of the form:

$$\left\{ \begin{pmatrix} 1 & x & y \\ 0 & 1 & z \\ 0 & 0 & 1 \end{pmatrix} : (x, y, z) \in \mathbb{R}^3 \right\}.$$

The usual matrix multiplication gives this set the structure of a group. Note that the only thing to be checked is the fact that this set is closed with respect to multiplication, and that every element in this set has an inverse that is also inside this set. Also, this group is clearly not commutative (the elements $(1, 0, 0)$ and $(0, 0, 1)$, for example, do not commute).

At the same time, notice that the underlying set is simply $\mathbb{R}^3$.

This group has a name: the Heisenberg group. It has had great historical importance in the developement of several subjects, such as Lie Group Theory, Harmonic Analysis, and Mechanics.

21. This group has $3^4 = 81$ elements.

A subgroup of order 9 is: $\begin{pmatrix} x & y \\ 0 & 0 \end{pmatrix}$.

A subgroup of order 27 is: $\begin{pmatrix} x & y \\ z & 0 \end{pmatrix}$.

A subgroup of order 3 is: $\begin{pmatrix} x & 0 \\ 0 & 0 \end{pmatrix}$.

23. Since ϕ is an isomorphism, it must satisfy $\phi(a \cdot b) = \phi(a) \cdot \phi(b)$. In particular, for $g \in G$, we must have:

$$\phi(g^k) = [\phi(g)]^k.$$

So, if $g^k = e_G$, $[\phi(g)]^k = \phi(e_G) = e_H$. In fact, we do not need ϕ to be an isomorphism. Homomorphism would have been enough.

26. The plane that is uniquely determined by p, q, and the origin intersects the sphere in a great circle. That is the great circle that we seek. The circle is unique as long as p and q are not antipodal (i.e., at opposite poles of the sphere). In that degenerate case, p, q, and the origin all lie on a line and do not determine a unique plane.

28. By composing with a map of the form ϕ_b, we may assume that τ maps the origin to the origin (i.e., we take p to be the origin 0). Then it is not difficult to see that τ must be a rotation. But we also know that τ fixes q. Clearly a rotation that fixes a point that is not 0 must be the identity.

29. Refer to Exercise 8.27 for terminology. We know that maps of the form $\rho_\theta \circ \phi_a$ take Poincaré lines to Poincaré lines. The simplest Poincaré line is an ordinary Euclidean diameter of the disc through the origin. It is of course determined by two points. It follows that all other Poincaré lines are determined by two points.

APPENDIX: Axiomatics

1. Using Proposition 5.4.3, we find:

$$
\begin{aligned}
\operatorname{card}\left[\left[S^T\right]^U\right] &= \left(\operatorname{card}\left[S\right]^T\right)^{\operatorname{card} U} \\
&= \left((\operatorname{card} S)^{\operatorname{card} T}\right)^{\operatorname{card} U} \\
&= (\operatorname{card} S)^{\operatorname{card} T \cdot \operatorname{card} U} \\
&= (\operatorname{card} S)^{\operatorname{card}[T \times U]} \\
&= \operatorname{card}\left(S^{T \times U}\right),
\end{aligned}
$$

since $\operatorname{card}(T \times S) = \operatorname{card} T \cdot \operatorname{card} S$ (c.f Exercise 3.23).

3. $\displaystyle\prod_{\alpha \in A} S_\alpha$ is defined as the set of functions

$$
f : A \to \left[\bigcup_{\alpha \in A} S_\alpha\right]
$$

such that

$$
f(\alpha) \in S_\alpha.
$$

This last condition is where the Axiom of Choice is implicitly used. The Axiom of Choice says that, given a collection of sets, we can construct a set by choosing one element from each of the sets in the collection. This is exactly what the condition $f(\alpha) \in S_\alpha$ requires.

5. Let A be a set, and apply the Axiom of Regularity to $\{A\}$. We see that there must be an element of $\{A\}$ which is disjoint from $\{A\}$. Since the only element of $\{A\}$ is A itself, we conclude that A is disjoint from $\{A\}$. Hence, since $A \in \{A\}$, we cannot have $A \in A$.

8. We say that A is a subset of B, and write $A \subset B$, precisely when the syllogism $x \in A \implies x \in B$ is true. This is an elementary statement in first-order logic, and does not run afoul of any of the Zermelo–Fraenkel axioms.

10. $S \cap T$ can be written as $\{x \in S \text{ such that } x \in T\}$. Now use the Axiom Schema of Replacement with $P(x,y)$ given by $P(x,y) = [(x = y) \wedge (y \in T)]$, and taking S as the set A that appears in the statement of the axiom. Then the hypotheses of the Axiom Schema of Replacement are clearly satisfied (because $(x = y) \wedge (x = z) \Rightarrow (x = z)$). So we must have that there is a set B so that $y \in B$ if $[(x \in S) \wedge (x = y) \wedge (y \in T)]$, which says precisely that B is the set $S \cap T$. Hence $S \cap T$ exists.

11. This is similar to the previous exercise: $S \setminus T$ can be written as $\{x \in S \text{ such that } x \notin T\}$. Now use the Axiom Schema of Replacement with $P(x,y)$ given by $P(x,y) = [(x = y) \wedge (y \notin T)]$, and taking S as the set A. Then the hypotheses of the Axiom Schema of Replacement are clearly satisfied (because $(x = y) \wedge (x = z) \Rightarrow (x = z)$). So we must have that there is a set B so that $y \in B$ if $[(x \in S) \wedge (x = y) \wedge (y \notin T)]$, which says precisely that B is the set $S \setminus T$. Hence $S \setminus T$ exists.

13. Use the Axiom Schema of Replacement, choosing

$$A = \{\varnothing, \{\varnothing\}\}$$

and

$$P(u,v) = (u = \varnothing \wedge v = x) \vee (u = \{\varnothing\} \wedge v = y).$$

The result follows immediately.

15. First, apply the Axiom of Pairing (see Exercise 5.13) with $x = A$ and $y = B$. This guarantees the existence of the set

$$C = \{A, B\}.$$

Consider the set $\cup C$, which is simply the set whose existence is guaranteed by the Sum Axiom. Note that:

$$x \in \cup C \, Si \exists D, (D \in C \wedge x \in D).$$

But such D can only be A or B (the only elements of C). Thus,

$$x \in \cup C \, Si (x \in A \vee x \in B).$$

By the Axiom of Replacement, we must have that the set whose existence is guaranteed by the Union Axiom is exactly the set

$$\cup\{A,B\},$$

which already existed before adding the Union Axiom. Thus, the Union Axiom is redundant.

17. If you have infinitely many pairs of shoes, then you can simply select the left shoe from each pair. There is no need for the Axiom of Choice. But with socks this strategy will not work (because you cannot tell a left sock from a right sock).

19. Suppose that $S \in T$ and $T \in S$. Apply the Axiom of Regularity to the set $\{S,T\} \neq \emptyset$ (which exists by the Axiom of Pairing—see Exercise 5.13). There must be an $x \in \{S,T\}$ such that $y \in x \Rightarrow y \notin \{S,T\}$. But the only elements of $\{S,T\}$ are S and T, so either $x = S$ or $x = T$. Suppose that $x = S$. If we had that $T \in S$, then we would have $T \in x$ and $T \in \{S,T\}$, which contradicts the choice of x above. Assuming now that $x = T$, the same problem arises. Thus we must have that our original supposition, $S \in T$ and $T \in S$, is impossible.

21. If $(a,c) \in A \times C$, then $a \in A$ and $c \in C$. But since $A \subset B$, $a \in B$, so $(a,c) \in B \times C$.

23. One can construct a bijection between the power set of X and the set of functions

$$f : X \to \{0,1\}.$$

To do this, let S be a subset of X. Let the function

$$f(t) = \begin{cases} 0 & \text{if} \quad t \notin S \\ 1 & \text{if} \quad t \in S \end{cases}$$

be associated to S. This gives a bijection between $\{0,1\}^X$ and $\mathcal{P}(X)$. When X is finite, one also has

$$\text{card}\,[\mathcal{P}(X)] = 2^{\text{card}\,[X]}.$$

25. Neither the rationals nor the reals is well ordered.

27. **(a)** The standard ordering ($<$) on the real number system.

   (b) The standard ordering ($<$) on the rational number system.

   (c) The so-called "lexicographic ordering" on ordered pairs of real numbers: We say that $(x,y) \triangle (x',y')$ if either **(i)** $x < x'$ or **(ii)** $x = x'$ and $y < y'$.

29. Clearly S is a set that can be described in fewer than 50 words, so it is certainly an element of itself. The axioms of set theory forbid

this set for the same reason that they forbid the set in Russell's paradox: a set can only be defined with reference to a *bona fide* set that already exists. For this set S, the "reference set" is the set of all sets, and that is not a legitimate set in our set theory.

Bibliography

[ADA]   J. F. Adams, On the non-existence of elements of Hopf invariant one, *Ann. of Math.* 72(1960), 20–104.

[AKS]   M. Agrawal, N. Kayal, and N. Saxena, PRIMES is in P, *Ann. of Mathematics* 160(2004), no. 2, 781–793.

[BAR]   J. Barwise, ed. *Handbook of Mathematical Logic*, North-Holland, Amsterdam, 1977.

[BLU]   M. Blum, "How to prove a theorem so no one else can claim it," *Proc. International Congress of Mathematicians* (Berkeley, CA, 1986), 1444–1451, AMS Providence, RI, 1987.

[BOM]   R. Bott and J. Milnor, On the parallelizability of the spheres, *Bull. Am. Math. Soc.* 64(1958), 87–89.

[BUS]   S. R. Buss, ed., *Handbook of Proof Theory*, Elsevier, Amsterdam, 1998.

[CHK]   C. C. Chang and J. Keisler, *Model Theory*, North-Holland, New York, 1977.

[COH]   P. J. Cohen, *Set Theory and the Continuum Hypothesis*, Benjamin, 1966.

[CUT]   N. Cutland, ed., *Nonstandard Analysis and its Applications*, Cambridge University Press, Cambridge, 1988.

[DAM1]   G. B. Dantzig and M. N. Thapa, *Linear programming 1: Introduction*, Springer-Verlag, New York, 1997.

[DAM2]   G. B. Dantzig and M. N. Thapa, *Linear Programming 2: Theory and Extensions*, Springer-Verlag, New York, 2003.

[DAV]   A. Davis, *Gödel's Theorem*, University of Oklahoma preprints, 1964.

[END]   H. Enderton, *A Mathematical Introduction to Logic*, 2nd ed., Academic Press, New York, 2001.

[ERD]   P. Erdős, On a new method in elementary number theory which leads to an elementary proof of the prime number theorem, *Proc. Nat. Acad. Sci. U. S. A.* 35(1949), 374-Ű384.

[FFP]   U. Feige, A. Fiat, and G. Persiano, "Noninteractive zero-knowledge proof systems," in *Advances in Cryptology*—CRYPTO '87 (Santa Barbara, CA, 1987), 52–72, *Lecture Notes in Computer Science* 293, Springer-Verlag, Berlin, 1988.

[FRA]   A. A. Frænkel, *Abstract Set Theory*, North-Holland Publishing, Amsterdam, 1966.

[GAM]   G. Gamow, *One Two Three . . . Infinity*, Dover Books, New York, 1988.

[GAJ]   M. R. Garey and D. S. Johnson, *Computers and Intractability: A Guide to the Theory of NP-Completeness*, W. H. Freeman and Co., San Francisco, 1991.

[GIH]   S. Givant and P. R. Halmos, *Logic as Algebra*, Mathematical Association of America, Washington, D.C., 1998.

[GRS]   R. L. Graham, B. L. Rothschild, and J. H. Spencer, *Ramsey Theory*, John Wiley & Sons, New York, 1980.

[GRE]   M. J. Greenberg, *Euclidean and Non-Euclidean Geometries*, 2nd ed., Freeman, San Francisco, 1980.

[HAL]   M. Hall, *The Theory of Groups*, Macmillan, New York, 1959.

[HALM]   P. R. Halmos, *Algebraic Logic*, Chelsea, New York, 1962.

[HER]   I. Herstein, *Topics in Algebra*, Xerox, Lexington, 1975.

[HJE]   K. Hrbacek and T. Jech, *Introduction to Set Theory*, 2nd ed., Dekker, New York, 1984.

[JEC]   T. Jech, *The Axiom of Choice*, North-Holland, New York, 1973.

[KAR]   R. Karp, The probabilistic analysis of some combinatorial search problems, *Algorithms and Complexity* (Proc. Sympos., Carnegie-Mellon Univ., Pittsburgh, Pa. 1976), 1–19; Academic Press, New York, 1976.

[KLE]   S. C. Kleene, *Introduction to Metamathematics*, Elsevier, New York, 1974.

[KRA1]   S. G. Krantz, *Real Analysis and Foundations*, CRC Press, Boca Raton, 1998.

[KRA2]   S. G. Krantz, *Essentials of Mathematical Thinking*, Taylor & Francis, Boca Raton, FL, 2017.

[KNU]   D. E. Knuth, *Surreal Numbers*, Addison-Wesley, Boston, MA, 1974.

[LAN]   S. Lang, *Algebra*, 3rd ed., Addison-Wesley, Reading, 1992.

[LIN]   T. Lindstrøm, An invitation to nonstandard analysis, in *Nonstandard Analysis and its Applications*, N. Cutland, ed., Cambridge Univ. Press, Cambridge, 1988.

[NAN]   E. Nagel and J. R. Newman, *Gödel's Proof*, New York University Press, New York, 1958.

[NEL]   E. Nelson, *Predicative Arithmetic*, Princeton University Press, Princeton, 1986.

[NIS]   N. Nissanke, *Introductory Logic and Sets for Computer Scientists*, Addison-Wesley, Reading, 1999.

[NSA]   http://members.tripod.com/PhilipApps/howto.html .

[RR1]   H. Rubin and J. Rubin, *Equivalents of the Axiom of Choice*, 1st ed., North-Holland, Amsterdam, 1963.

[RR2]   H. Rubin and J. Rubin, *Equivalents of the Axiom of Choice*, 2nd ed., North-Holland, Amsterdam, 1985.

[RUD]   W. Rudin, *Principles of Mathematical Analysis*, 3rd ed., McGraw-Hill, New York, 1976.

[SCL]   C. P. Schnorr and H. W. Lenstra, Jr., A Monte Carlo factoring algorithm with linear storage, *Math. Comput.* 43(1984), 289–311,

[SEL]   A. Selberg, An elementary proof of the prime-number theorem, *Ann. of Math.* 50(1949), 305Ü313.

[SCH]   J. R. Shoenfield, *Mathematical Logic*, Addison-Wesley, Reading, 1967.

[SIN]   S. Singh and J. Lynch, *Fermat's Enigma: The Epic Quest to Solve the World's Greatest Mathematical Problem*, Anchor Books, New York, 1998.

[SMU]   R. Smullyan, *Gödel's Incompleteness Theorems*, Oxford University Press, New York, 1992.

[STO]   R. R. Stoll, *Sets, Logic, and Axiomatic Theories*, W. H. Freeman and Company, San Francisco, 1968.

[SUP]   P. Suppes, *Axiomatic Set Theory*, Dover Publications, New York, 1972.

[WHB]   A. N. Whitehead and B. Russell, *Principia Mathematica*, 2nd ed., Cambridge Univ. Press, Cambridge, 1950.

Index

Abel, N., 198, 200
abelian group, 198
absolute value, 131, 155
absurd numbers, 114
accumulation point, 158
Adams, J., 146
addition, 122
    associativity of, 119, 122
    closure of, 122
    commutativity of, 119, 122
    of natural numbers, 113
additive
    identity, 122
    inverse, 122
    inverse in the integers, 119
adjacency matrix, 191, 192
Adleman, L., 184
airline scheduling, 186
algebraic
    numbers, 94
    numbers are countable, 96
algebraically complete, 143
Algorithm for Finding the Greastest
    Common Divisor, 173
analogy
    reasoning by, 35
and, 3, 13
anti-symmetry, 77
Archimedean Property, 129
    of the real numbers, 129
Aristotelian logic, 7
assign, 79, 80
associative, 197
atomic statement, 1
Axiom for Cardinals, 230, 231
Axiom of Choice, 135, 230, 231, 233,
    234

examples, 235
Axiom of Extensionality, 229, 230
Axiom of Infinity, 230, 231
Axiom of Regularity, 229, 230
Axiom of Replacement, 231
Axiom of Substitution, 18
Axiom Schema of Replacement, 18,
    230
axioms, 2
    for a field, 122
    for arithmetic, 110
    for the natural numbers, 110
    of a group, 197
    of Euclid's geometry, 237
    of geometry, 211
    of set theory, 229

balance scale, 54
Banach–Tarski paradox, 236
basis for a vector space, 235
Beltrami–Klein model of
    non-Euclidean geometry,
    217
bijection, 86, 90, 91, 98
binary operation, 197
Bolyai, F., 212
Bolyai, J., 212, 214, 239
bounded above, 127
Brouwer Fixed Point Theorem, 130
Brouwer Fixed-Point Theorem, 130

Caesar, Julius, 183
cancellation, 115
Cantor set, 160
    is uncountable, 162
    length of, 161
    uncountability of, 162

287

Printed in the United States
by Baker & Taylor Publisher Services

Printed in the United States
by Baker & Taylor Publisher Services